A Geoscience Guide to the Burgess Shale

Geology and Paleontology in Yoho National Park

Murray Coppold
and
Wayne Powell

The Burgess Shale Geoscience Foundation

Second Edition

P.O. Box 148, Field, British Columbia, Canada V0A 1G0
Telephone: 250 343 6006
info@burgess-shale.bc.ca
www.burgess-shale.bc.ca

Design/Production
Murray Coppold

International Standard Book Number (ISBN) 0-9780132-0-4
Printed in Canada

First printing 2006
Second printing 2014

Contents

During the field season of 1910 the study of the Cambrian strata of the section of the Rocky Mountains adjacent to the main line of the Canadian Pacific Railway was continued and special attention given to the Stephen formation. Its outcrop was carefully examined for many miles along the mountain sides with the hope of finding a locality where conditions had been favorable for ... preservation of examples of much of the life that ... must have existed in the Middle Cambrian seas.

The finding, during the season of 1909, of a block of fossiliferous siliceous shale that had been brought down by a snow slide on the slope between Mount Field and Mount Wapta, led us to make a thorough examination of the section above in 1910. Accompanied by my two sons, Sidney and Stuart, every layer of limestone and shale above was examined until we finally located the fossil-bearing band. After that, for thirty days we quarried the shale, slid it down the mountain side in blocks to a trail, and transported it to camp on pack horses, where, assisted by Mrs. Walcott, the shale was split, trimmed, and packed, and then taken down to the railway station at Field, 3000 feet below.

—Charles Doolittle Walcott, 1911

Yoho National Park

In 1884 the Canadian Pacific Railway reached the Kicking Horse Valley where it established the town of Field. The railway opened the valley to tourists, adventurers and scientists, including R.G. McConnell, a geologist with the Geological Survey of Canada. In early September 1886, McConnell climbed the flank of Mt. Stephen where he discovered a rich fossil bed 520 metres (1700 feet) above Field.

In 1886, the Government of Canada established the young country's second national park, only twenty eight years after the first European, James Hector, laid eyes on the area. The tiny park encompassed only 26 square kilometres at the base of Mt. Stephen. In 1901 the park was expanded to include today's area plus the Beaverfoot Valley to the west. Yoho National Park changed size three more times before the boundaries were finally set in 1930, enclosing 1,313 square kilometres.

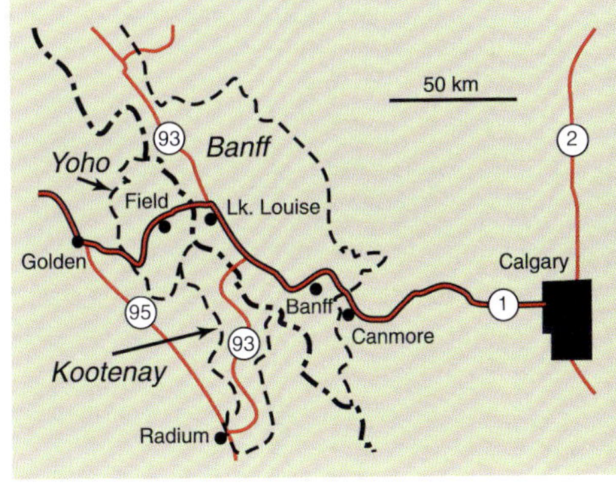

In the early 1900s, Charles Walcott came to study Yoho's trilobites. In 1909, while searching for other fossil sites, Walcott discovered the well-preserved soft-bodied fossils of the Burgess Shale on the ridge between Wapta Mountain and Mt. Field.

Industry had been quick to begin exploiting the rich natural resources of the Yoho area. Pre-existing logging and mining activities were allowed to continue after the park expanded and encompassed their operations. However, no new permits were issued. Mining ceased in 1952, logging in 1968.

Today, the value of Yoho National Park lies in its conservation. By reading the messages in the mountains, the climate record and the fossils, the visitor can reflect upon the history of our planet and come away with a heightened sense of awareness.

The Meaning of World Heritage

If not for the protected status of the national park, the Burgess Shale fossils would have been plundered long ago by commercial quarrying. The protection they have been afforded made it possible for the 1975 Royal Ontario Museum expedition to find, still lying in the talus, the counterpart (other half) of a rare *Branchiocaris* specimen collected by Percy Raymond forty-five years previously!

Takakkaw Falls in the Yoho Valley

The Convention Concerning the Protection of the World Cultural and Natural Heritage was adopted by the United Nations Educational, Scientific and Cultural Organization (UNESCO) in 1972. The convention's objective is to recognize sites of global significance, which should be held in trust by nations for humankind, and to ensure support of the international community for their preservation. Although over 112 countries are signatories to the convention, World Heritage status is given to relatively few outstanding protected areas.

In 1984, UNESCO declared Yoho, Banff, Kootenay and Jasper National Parks, together with Mt. Robson, Hamber and Mt. Assiniboine Provincial Parks, as the Rocky Mountain World Heritage Parks. Within this designation lies the Burgess Shales World Heritage Site—a protected area within a protected area.

Canada, as steward of the Rocky Mountain World Heritage Parks, has obligations to UNESCO for their protection. Thus access to the Burgess Shale is restricted to those accompanied by a qualified, licensed guide, and collection of fossils is prohibited.

About Time

Our understanding of geological time has come a long way since the Irish archbishop James Ussher (d. 1656) calculated from studies of the Scriptures that Earth was formed in 4004 B.C. The current estimate of the age of Earth is 4.5 billion years.

So much has happened to the rocks on Earth since it was born that it is impossible to determine the age of our planet from Earth rocks. (The oldest rocks yet discovered on Earth are approximately four billion years old, and are found in Canada's Northwest Territories.) Instead, geoscientists indirectly determined the Earth's age by examining the decay of radioactive isotopes in meteorites — rocks that had been floating undisturbed in space since the solar system formed.

Estimating geologic time was a real problem before the discovery of radiometric dating. Geologists in Charles Walcott's day attempted to measure time by averaging the rates of natural processes — how long it took Earth to cool from an initial molten state, how long for thick sequences of carbonate rocks to be deposited, how long for the initially-fresh oceans to become salty as minerals were washed in from continental weathering, even how long the Sun could have burned.

Charles Darwin had assumed almost limitless time, in order to account for evolution. In his first edition of *The Origin of Species* (1859), Darwin suggested 300 million years had elapsed since the late Mesozoic — a figure we know to be too high, but one of the first indications of the immensity of time. Most other late nineteenth century estimates for the age of Earth were less than 100 million years, causing Darwin to abandon his number in later editions of *The Origin*.

In 1893 Charles Walcott tried his hand at using estimates of erosion and deposition rates to calculate the length of time encompassed by thick Paleozoic rocks of the Western Cordillera. Scaling-up his number to cover the Mesozoic and Cenozoic, and adding in a Precambrian estimate, he came up with about 55 million years. The Cambrian sat mid-way in the range, no doubt intrinsically pleasing to scientists of the day as they contemplated the appearance of life on Earth.

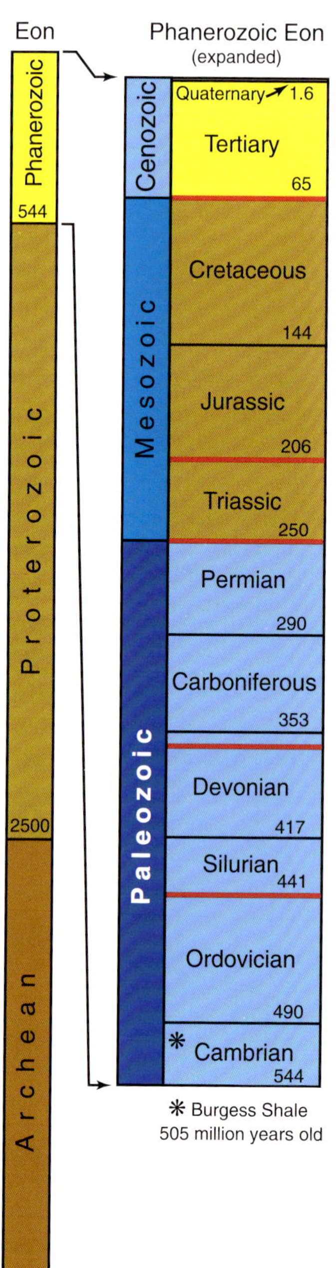

* Burgess Shale
505 million years old

The chart at left shows the subdivisions of geologic time, with the bounding ages indicated in millions of years.

At far left are the three eons since the formation of Earth more than four billion years ago. Phanerozoic means *visible life*, an obvious reference to the macrofossils readily seen in its rocks. The Proterozoic and Archean comprise the preceding years, loosely termed the Precambrian. Proterozoic means *earlier life* and Archean means *ancient*. With amazing prescience, these names were assigned before most of the microscopic single-celled fossils characteristic of these eons were discovered.

At near left, the Phanerozoic Eon is expanded to show its eras (Paleozoic, Mesozoic and Cenozoic) and periods. The Phanerozoic periods are subdivided into hundreds of stages, which are defined by evolutionary changes in fossil assemblages contained within the rocks.

The times of five major crises in the history of life, commonly termed *mass extinctions*, are shown by the red lines. Here we get into a circular argument. Since the periods are defined by fossil assemblages, it is not surprising that the extinctions occur at period boundaries.

Compressing Time

If you have difficulty conceiving of a million years, much less more than four *billion*, try imagining the 4.5 billion years of geologic time compressed into one year. Each day would represent a little more than 12 million years and the following significant dates would occur:

Formation of planet Earth	January 1
Meteor bombardment ends	February 17
Simple cellular life appears	March 20
Atmospheric oxygen becomes abundant	July 22
Ediacara fauna appears	November 17
Burgess Shale fauna appears	November 20
First land plants	November 28
Permian extinction	December 13
Dinosaurs present	December 21-27
End-Cretaceous extinction	December 27
Pleistocene glaciations begin	December 31, 8:45 PM
Ancestral Man (*Homo*)	December 31, 9:07 PM
Man (*Homo sapiens*)	December 31, 11:45 PM

The Rise of the Rockies

Throughout its history the configuration of Earth's oceans and continents has changed constantly, albeit slowly. The plates that comprise the crust all move relative to each other, their movements driven by convection cells in the underlying mantle. Some grow; some are consumed. In this process the shape and size of oceans and continents change over time. Since it first began to open 200 million years ago the Atlantic Ocean has been expanding, at the expense of the shrinking Pacific Ocean. The closing of the Pacific Ocean has shaped western North America, from Vancouver Island through to the Alberta plains, and continues to do so today.

Plate Tectonics: The Push Behind the Mountains

Driven by mantle convection cells, the Pacific sea floor spreads. Where it encounters a more buoyant continental mass, such as the North American continent, the ocean floor is subducted (driven down) beneath the continent and re-melted as it descends back into the mantle. Compression is created where a continent overrides an oceanic plate, deforming continental rocks into mountain ranges.

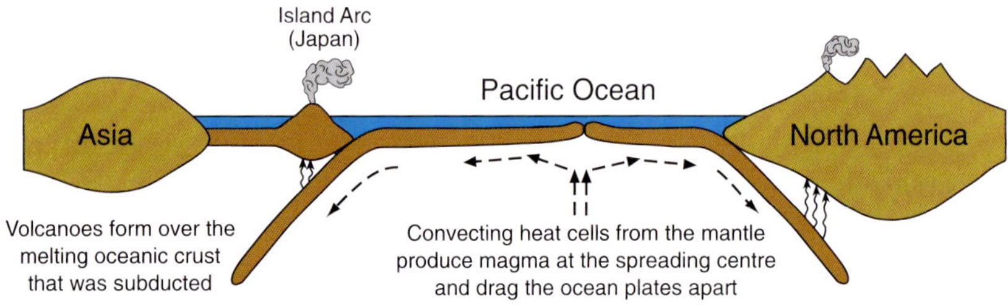

This process builds mountains in two different ways. Volcanoes form over the region where the subducted rock slab begins to melt. Some volcanoes related to subduction form on the continent, such as Mt. Garibaldi or Mt. St. Helens. Others form long, arcuate volcanic islands akin to Japan, which are later accreted to the edge of the continent. The drag associated with the subducting crust exerts compression on the continental edge. This compression shears and fractures rocks of the margin, pushing them up into mountain ranges.

You can imagine what happens when such enormous masses of rock collide by thinking of the last time you shovelled snow from your driveway. Think of the snow shovel as the volcanic island chain riding toward the continent. When the shovel first hits the snow and pushes it forward, the snow immediately in front of the shovel begins to thicken and bulge: the pressure from the shovel forces the snow to compress by folding the layers. As you continue to push the shovel forward, the cohesiveness of the snow is exceeded, and it breaks into thin sheets that ride forward and curl up. Instead of folding, thrust faults have begun to form in the snow.

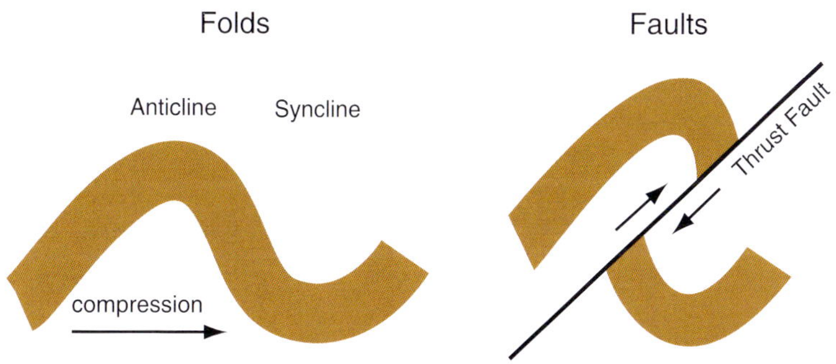

Rocks behave similarly. When enormous stresses are placed on them the strata will start to bend. Rocks buried at depth in the earth are very hot, and therefore somewhat plastic. They will bend until they form tight folds. Rocks closer to the surface are cool, and are more brittle; they will snap after they have been bent beyond a certain point. The most common faults that form during mountain building are called thrust faults, and have the characteristic feature of stacking older rock on top of younger rock. Thick slabs of rock slide forward and stack, one upon another, along these shallow-angle faults, resulting in a shorter but thicker (that is, compressed) stack of rock.

Building the Rockies

The mountains of western Canada started to form 175 million years ago, during the Middle Jurassic Period. At that time the first volcanic island complex (called the Intermontane Superterrane by geologists)

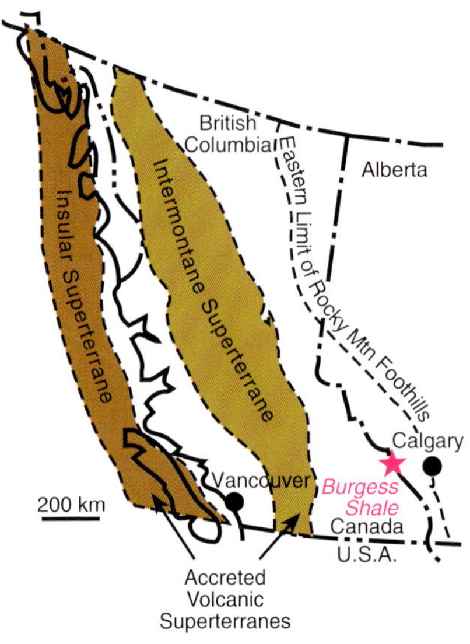

collided with the ancient margin of North America. Most of this enormous island slid beneath the continent but part was cropped-off and added to the edge of North America. These volcanic rocks underlie a north-south strip of central British Columbia, including the Nelson-Kamloops area in the southern end of the province.

The Intermontane Superterrane acted like an enormous bulldozer along the continental margin. As the volcanic landmass accreted onto the North American plate, it pushed the sedimentary rocks in front of it eastward, folding and faulting them in the process. As the deformation reached the region that we know as the Rockies, the peaks of the Main Ranges rose from the ocean.

A second volcanic island (known as the Insular Superterrane) struck North America during the Late Cretaceous, 85 million years ago. The accretion of this landmass, which now underlies Vancouver Island,

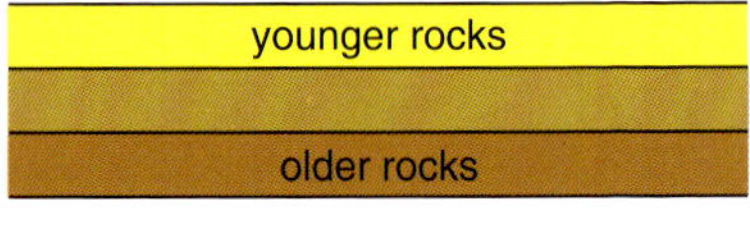

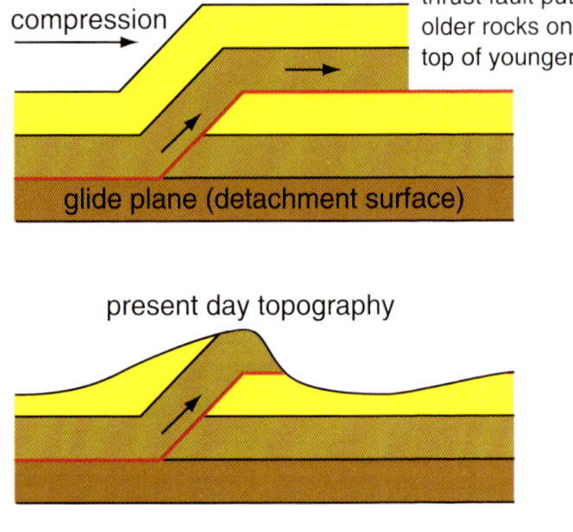

Thrust faulting.
Layered sedimentary rocks are deposited with younger rocks on top of older (upper diagram). The compression associated with continental drift buckles the layers, causing faults to form along planes of weakness (red line). Older rocks are thrust above younger (middle). Structure and subsequent erosion results in east-facing scarps and gentler west-facing dip slopes characteristic of the Rocky Mountain front ranges.

affected rocks all the way to Calgary. It was during this second mountain building event that the easternmost peaks of the Rockies grew and the rocks below the foothills crumpled.

By the Eocene Period (45 million years ago), the Insular Superterrane stopped its eastward push. Without this continued compression from offshore, folds and thrust faults stopped forming. Now a new set of faults developed. Unlike the compressional thrust faults that built the mountains, these new normal (extensional) faults acted to relieve some of the stresses of uplift. Huge blocks of rock dropped between sets of north-south faults. Mt. Field and Mt. Stephen, lying between the Stephen-Cathedral and Fossil Gully normal faults, are examples of such down-dropped blocks.

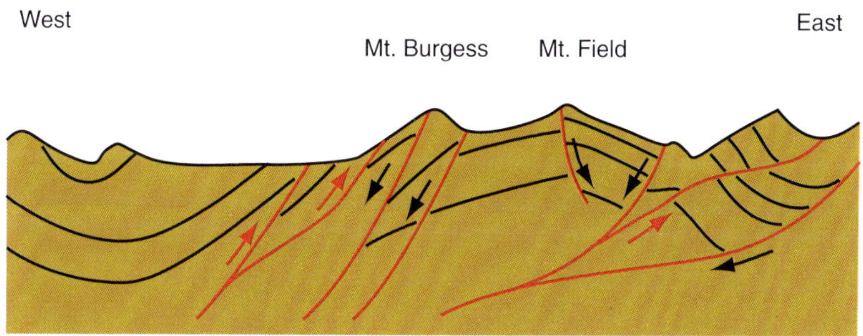

In contrast to block models, a complex pattern of folds and faults is created by mountain building. Thrust faults (red arrows) pre-date normal faults (black arrows). It took many years of detailed mapping to understand the structural relationships in Yoho National Park. Even this diagram is simplified.

Ranges of the Rockies

The Rocky Mountains are the easternmost slice of the North American Cordillera, the broad band of mountains that run along western North America. The central Rockies span the distance from the foothills west of Calgary to the Rocky Mountain Trench (the towns of Radium and Golden lie in the Trench) and can be divided into bands based on the shape and style of the mountains. Changes in the shape of mountains are directly related to changes in the underlying geology.

The Front Ranges rise dramatically from the low-lying foothills. Front Range mountains are typified by Mt. Rundle in Banff with its wave-like shape. The eastern face is steep, whereas the western slope rises

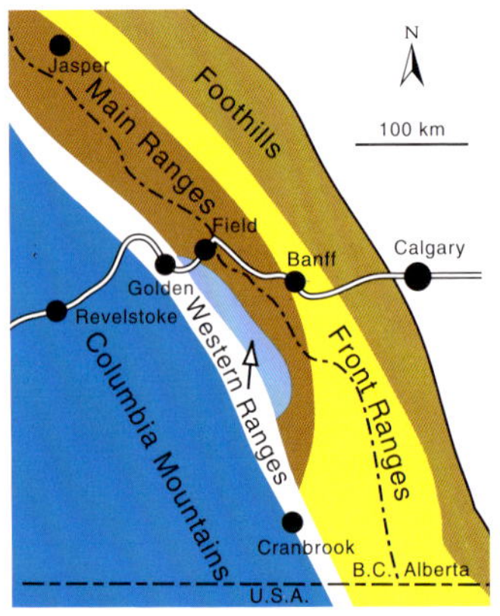

Ranges of the Rockies

much more gradually. The mountain takes this shape because the alternating layers of limestone and shale are tilted up to the east. The eastern slopes correspond to the resistant limestone strata. The shales erode leaving a valley between the mountain ridges.

The Main Ranges lie west of Banff. Main Range mountains are typified by Castle Mountain or Cathedral Mountain. These mountains take on a fortress-like appearance with their high cliff walls and flat tops that are often decorated with spires. The rock strata underlying the Main Ranges are close to flat-lying, producing their flat tops. The cliffs result from the erosion style of the limestone layers.

The Western Ranges have sharp-peaked mountains. In this region the rock layers are tightly folded and are almost standing on end. Often the strata have been folded back on themselves. The mountains are steep-sloped and peaked because of the steep, folded nature of the rock strata.

Erosion at Work

Erosion is another powerful force of nature that has been acting to level the mountains of Western Canada. Wind and water began wearing down the mountains since the moment the rocks first rose from the ocean. Together with the power of the grinding glaciers that repeatedly advanced across the Rockies over the past 240,000 years, these elements have worn down and shaped today's mountain landscape.

Flowing water is the main force of erosion in the Rockies today. Each day mountain streams carry away glacial debris and freshly weathered rock from the high slopes. The evidence is in the water all around us—the Kicking Horse River flows white with silt, and the blue-green mountain lakes derive their colour from sunlight diffracted by glacier-ground clay particles.

Up In The Mine

A mine. In a National Park. Inside a mountain. Not what you would expect, but for over sixty years zinc and lead were mined from within Mt. Stephen and Mt. Field as the only successful metal mines in the Canadian Rockies.

Railway construction workers chanced upon lead-zinc ore in the talus from Mt. Stephen, but pioneer guide Tom Wilson was first to stake the ground in 1882. The East Monarch deposit averaged 7% lead, 10% zinc and 1.2 ounces of silver per ton, but it was not economically viable until the 1917 discovery of the richer West Monarch ore body—only 250 metres away, but on an inaccessible cliff face

The ore was brought around a corner from West to East Monarch on a crude tramway supported by steel brackets affixed to the cliff. From there it was transported underground to the base of the cliff.

Full scale activity at the Monarch Mine began in 1928, when Base Metals Mining Corporation built a camp and mill at the base of Mt. Stephen by Monarch Creek, and an aerial tram on which miners rode up to work. An aerial tram to the Kicking Horse mine was installed in 1937, and both mines were active from 1935 until 1952. In all, ore production exceeded 600,000 tons.

Within the East and West Monarch mines, stopes extend back over 700 metres into the mountain. The Kicking Horse stopes have a similar extent

Tough to find. Timber shoring on a Mt. Stephen portal.

into Mt. Field. Yet in spite of all the activity, miners made only a few small openings in the cliff faces. Looking north from the Monarch Campground, the Kicking Horse Mine portals are visible just above the talus slope on Mt. Field. The Monarch Mine portals on Mt. Stephen are on a ledge separating darker limestones below from light-coloured dolomite above, near the base of the Cathedral Formation.

The Cambrian World

The Cambrian world map looked very different to that of today. For perhaps half a billion years, all the world's landmasses had been clustered in the Southern Hemisphere as the supercontinent Rodinia. In the late Proterozoic, Rodinia broke up and the continents drifted apart briefly only to coalesce again as the supercontinent of Pannotia. About 650 million years ago, most of the landmasses in temperate and high latitudes experienced an ice age more severe, and of much longer duration, than the Pleistocene.

In the early Cambrian, ancestral North America — *Laurentia* — broke free of Pannotia and drifted in the Panthalassic ("all sea") Ocean to the equator. In so doing, it created an intervening sea, the Iapetus Ocean, a Paleozoic version of today's Atlantic.

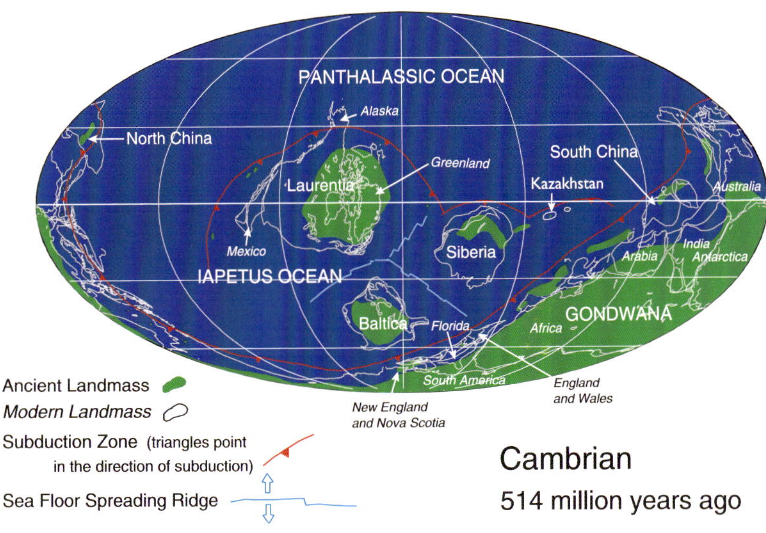

Things were different in the Cambrian. Earth revolved faster — the days were 21 hours long and there were 420 of them in a year. The Moon was closer to Earth and consequently, tides were stronger. The Sun's brightness was weaker than it is today, but a higher concentration of carbon dioxide in the atmosphere compensated for the decreased solar energy and maintained temperatures acceptable for life.

The skies were empty of birds; there were no fish in the seas. There were no land plants, and without a protective covering, continental erosion rates were high.

Laurentia remained in tropical latitudes for the next 300 million years. The continental landmasses, being less dense, have persisted through time. In contrast, none of the pre-Jurassic ocean basins remain. The more dense oceanic lithosphere has long since been subducted back into the mantle by the relentless conveyor belt of plate tectonics.

Tropical Fossils

Without question the Burgess Shale is the most famous Cambrian fossil bed, but it is not alone. Paleontologists have found soft-bodied animal fossils in Cambrian rocks at over thirty sites around the world, including Greenland, China, Siberia, and Australia.

Each new locality has preserved several of its own unique animals, but more importantly, many of the unusual animals of the Burgess Shale are found in the other Cambrian fossil beds. This proves that the Burgess Shale fossils are not some freak occurrence, but rather represent animals that inhabited widely dispersed regions of Earth's ancient oceans. They are probably typical of the Cambrian low-latitude fauna — the classic localities in Canada, Greenland, China, Siberia, and Australia all lay in the Cambrian tropics.

The Cambrian fauna developed in wide, shallow, warm seas. Biological productivity was high, owing to the abundant nutrients brought in by continental weathering and oceanic upwelling on the continental margins. The stage was set for the Cambrian explosion.

At The Edge of An Ancient Continent

Those of you not wearing a bathing suit and a life preserver would be ill prepared if you somehow found yourselves in Yoho National Park 505 million years ago. During the Middle Cambrian the Burgess Shale quarries lay just north of the equator, far offshore from the ancient coast of North America. Sediments deposited on the bottom of this ancient tropical ocean formed all the rocks that we see across the Kicking Horse, Emerald and Yoho Valleys.

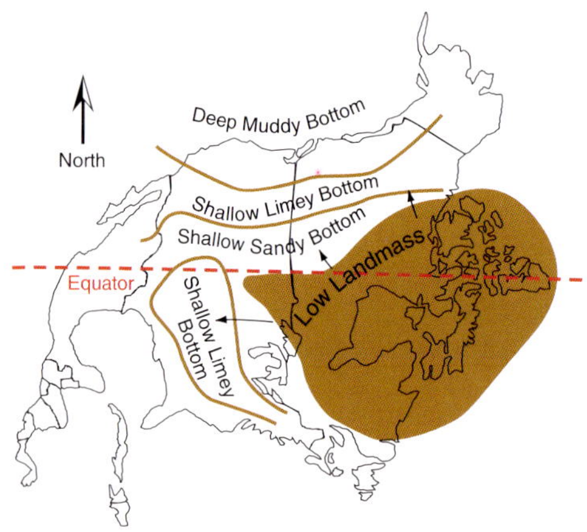

In the Cambrian, ancestral North America was turned at a right angle to its present orientation, and straddled the equator. The asterisk in western Canada marks the position of the Burgess Shale.

Beginning in the Early Cambrian Period (~544 million years ago) sea level rose and flooded coastal regions of ancient North America. Land plants had not yet evolved, and erosion rates were high on the vegetation-free continent, reducing the landmass to low relief. Vast deposits of beach sand were the first Cambrian sediments laid down on the eroded Precambrian bedrock. Tides were more extreme in the Cambrian, and the tidal cycles worked and reworked the beach sediments, winnowing out all but the most resistant minerals. The final result was the spectacular sandstones of the Gog Group, over two kilometres thick, composed almost entirely of quartz. Although these sandstones are largely covered in the Kicking Horse Valley, most of the popular hikes in the Lake Louise and Moraine Lake areas are almost entirely on rocks of the Gog Group.

The rest of the Cambrian saw several cycles of sea level rise and fall. With a rise in sea level, the shoreline was pushed back several hundred miles over the low-lying continent, to what is now central Canada. With reduced sediment input from erosion, limestone deposition dominated the rest of the Cambrian. A shallow carbonate shelf, similar to today's Bahama Banks, covered Alberta and most of Saskatchewan. Mudstone continued to be deposited in the deep ocean (present day British Columbia), and at times of high sea level was deposited over the shelf as well.

Geologists subdivide sedimentary rocks into packages that can be mapped regionally. These basic rock units are called formations. Alternating mudstone and shallow-water limestone comprise all of the Middle Cambrian sedimentary formations of eastern Yoho National Park. The formations are characterized by cycles of deposition which alternated between shaly rocks in the lower portion and cleaner carbonates above. Starting at the base, we have the Mt. Whyte-Cathedral, the Stephen-Eldon, and the lower Pika-upper Pika cycles. The position of the Burgess Shale fossils in the Burgess Shale Formation is indicated in the diagram below by the letter *F*.

The Mt. Whyte, Stephen, and basal Pika half-cycles are dominantly subtidal to locally intertidal mudstones and siltstones. The Cathedral, Eldon, and upper Pika half-cycles are all composed of shallow-marine carbonates. In the carbonate half-cycles, algae were important sediment contributors and binders near the seaward edge. Behind the algal rim, subtidal carbonate mudstones are the main rock type.

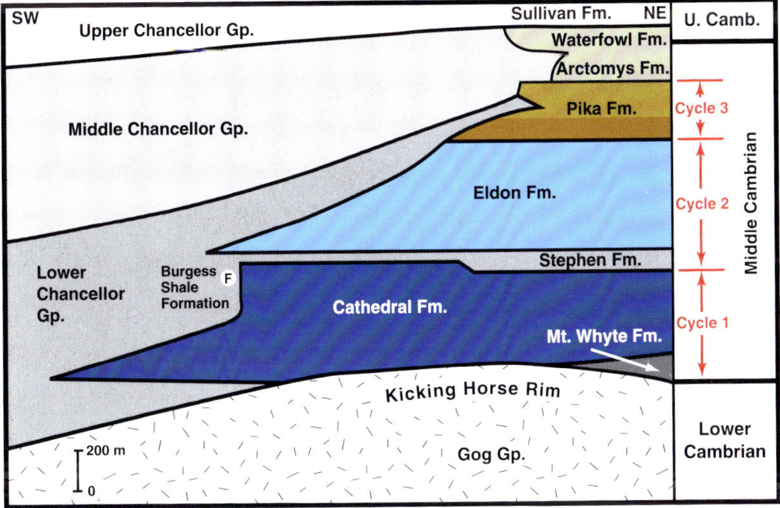

Topographic Map of the Kicking Horse Pass Area

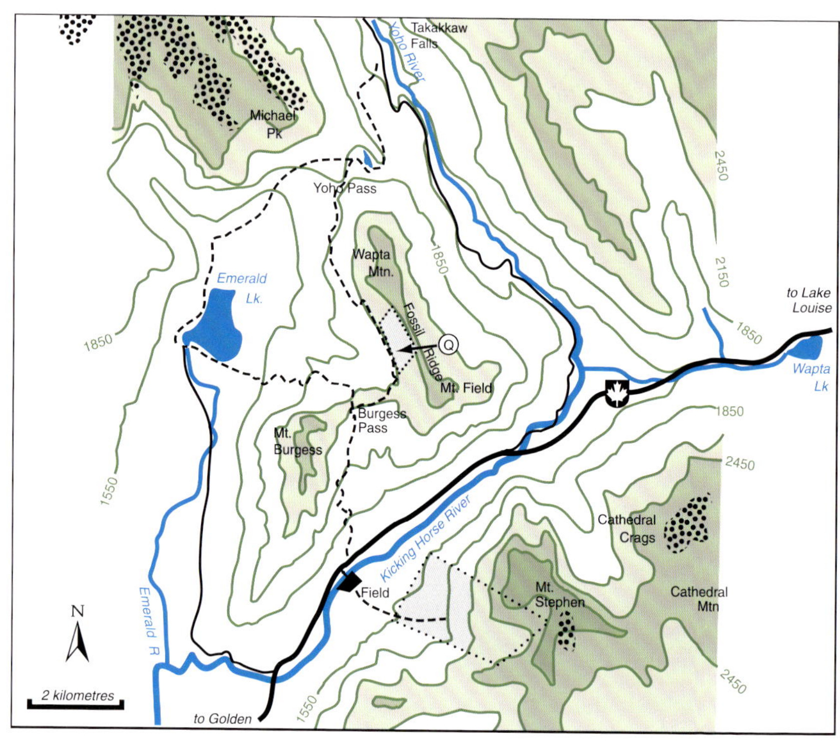

Topographic setting of the fossil localities. Contours shown in metres, with an interval of 300 metres or approximately 1,000 feet elevation. Higher elevations are shaded. The Burgess Shale quarries lie at the location marked 'Q' in the centre of the map. Note Parks Canada Restricted Access zones covering the Burgess Shale and the Mt. Stephen Fossil Beds.

Elevation (interval 300 m or 1000 ft)

Ice

Rivers

Trails

Parks Canada Restricted Access Area

Burgess Shale Quarries Q

Trans-Canada Highway

Geologic Map of the Middle Cambrian Formations

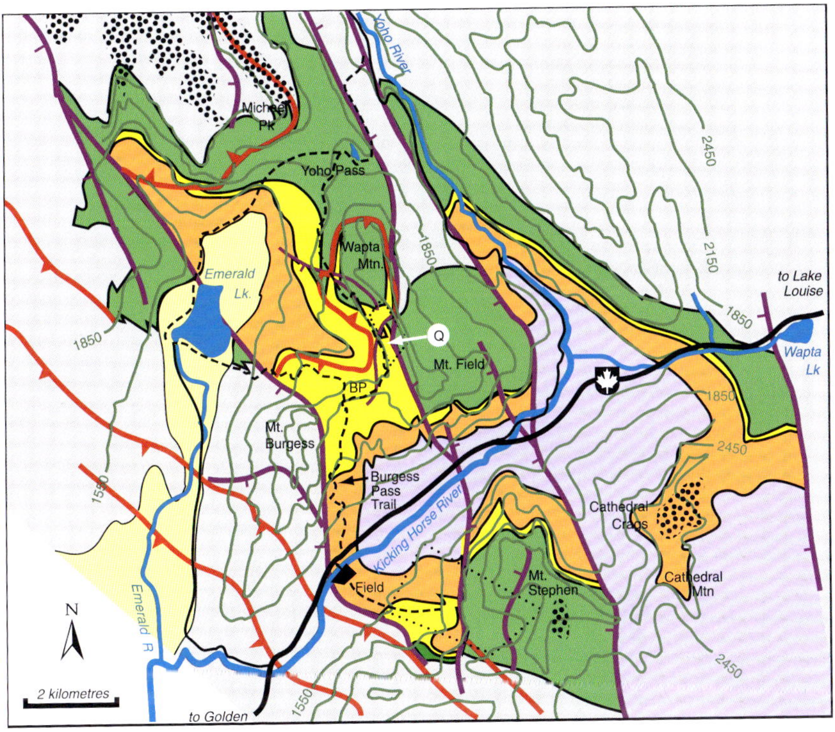

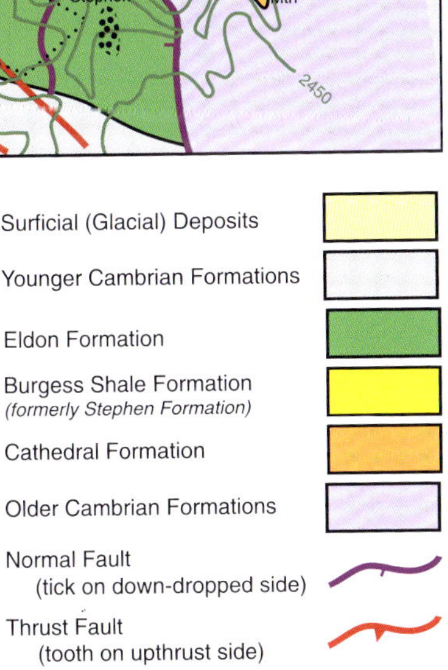

Generalized geologic map showing the Middle Cambrian Cathedral, Burgess Shale and Eldon Formations (see stratigraphic chart, page 15). The Burgess Shale Formation was formerly referred to as "thick Stephen Formation" until it was re-defined in 1998. Rocks to the west are Middle and Upper Cambrian Chancellor Group shales and slates. To the north and east are Upper Cambrian carbonate rocks, while to the southeast lie older Cambrian Gog and Mt. Whyte sandstones. Maps after Price and others (1980) and Balkwill and others (1980).

The Middle Cambrian finished with the Arctomys Formation being deposited in very shallow water. It exhibits mud cracks and casts of halite crystals, attesting to restricted water circulation and periodic drying of the sediments.

The front of each successive limestone unit lies almost on top of each of the others. Because the break between deep and shallow water environments repeatedly re-developed in the same place, this edge of the shallow water rim of Cambrian North America is referred to as the Kicking Horse Rim. All the rocks to the west are composed of shales and deep water carbonates of the Chancellor Group.

The Cathedral Escarpment

A spectacular submarine cliff over 100 metres high — the Cathedral Escarpment — marked the front of the algal limestones of the Cathedral Formation. The Cathedral Escarpment was initially thought to be a product of prolific algal growth and sediment accumulation during a period of sea level rise. A recent re-interpretation suggests that the escarpment is the headwall of a regional slide scar, where the over-steepened front of the Cathedral Formation collapsed into the basin shales. In either case, the Cathedral Formation carbonates are older than the adjacent mudstones of the Burgess Shale Formation.

Over approximately two million years, the Burgess Shale Formation mudstones filled the basin in front of the escarpment, and eventually covered the Cathedral limestone when sea level rose. The Stephen Formation mudstones are much thinner on top of the escarpment than the Burgess Shale mudstones in front of it. The Cathedral Escarpment marks this change in thickness and it can be traced across much of Yoho National Park.

All of the soft-bodied fossil localities in Yoho Park, and there are more than a dozen, lie in the thick Burgess Shale Formation immediately beside the Cathedral Escarpment. As we will see in the next chapter, this submarine cliff was crucial to fossil formation and preservation in the Burgess Shale: no Cathedral Escarpment, no fossils.

Evolution and the Burgess Shale

Life appeared on Earth relatively quickly, geologically speaking. Primitive life inhabited Earth's ocean within one billion years of our planet's coalescence. In 1982, paleontologists discovered the oldest fossils yet found in 3.4 billion year old chert beds in northwestern Australia (Warrawoona Group). These microfossils, less than 0.02 millimetres in diameter, are the preserved forms of bacteria and cyanobacteria (primitive photosynthetic organisms). Together, they comprise the most primitive kingdom of life: single-celled organisms in which DNA floats freely within the cell. Respiration and photosynthesis take place directly through the cell wall. They are known as prokaryotes (meaning *pre-nucleus*).

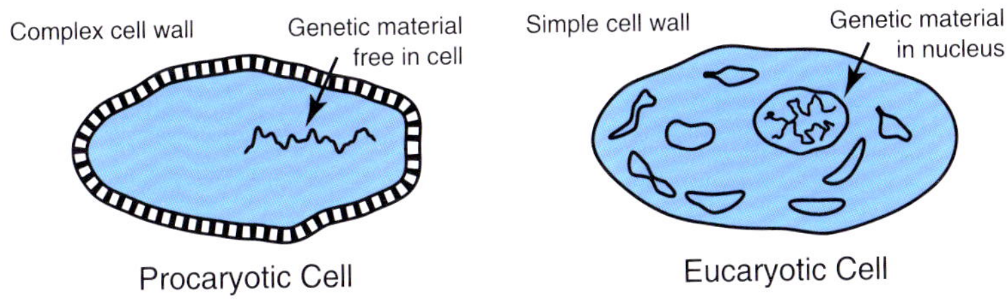

Bacteria and cyanobacteria remained the only life forms until the first eucaryotic organisms appeared over two billion years ago. Eucaryotes contain a nucleus which encapsulates the organism's DNA within the cell, and the term means *true nucleus*. Respiration and photosynthesis are handled by specialized organelles within the cell. These newly evolved algal plankton may have evolved from groups of procaryotes that initially lived together symbiotically.

Unlike procaryotes, these more complex single-celled organisms have their DNA organized into chromosomes. They evolved from reproducing asexually (simple splitting of the parent cell to produce clones), to sexual reproduction. In sexual reproduction, specialized cells from different organisms unite to form a new individual. Increased genetic variation may explain the rapid increase in the rate of evolutionary change after the appearance of eucaryotes.

Where are the Bodies?

The next step in the evolution of life on Earth was the development of multicellular life — plants and animals. Algae grouped together to form primitive seaweed nearly two billion years ago. The first animals followed later, but these creatures must have been entirely soft-bodied because no skeletal traces have been found. Instead, paleontologists find their fossilized burrows and tracks, called trace fossils. The oldest trace fossils are approximately 750 million years old, but trace fossils did not become abundant for almost another 100 million years.

The beauty of geology is that it is truly a multi-disciplinary science. In the absence of body fossils, molecular chemists have stepped in to look for biomarkers — certain carbon molecules produced by life processes — in Precambrian rocks. The oldest known biomarkers, reported in 1999, indicate both prokaryotes and eukaryotes were present 2.7 billion years ago. Biologists can estimate the origin of animal groups by looking at current rates of gene sequence divergence in a wide variety of organisms. If these rates have been constant through time, extrapolating backward estimates the time of origin of the major phyla. The assumptions involved are controversial, but estimates range from about 600 to 1,000 million years ago — long before body fossils appear in the rock record.

Impressions of possible primitive animals are found in rocks that are as old as 680 million years. These Ediacaran fossils, named after the town of Ediacara in South Australia, are preserved as imprints in sandstone that was deposited just after a global ice age. Commonly disk- or feather-shaped, these fossils have traditionally been viewed as early sea-pens and marine worms. However, some paleontologists disagree, believing that the Ediacaran fauna are a failed evolutionary experiment and are unrelated to any modern animals. Geologists have discovered several Ediacaran fossil sites in western Canada, including a site in Mt. Robson Provincial Park, approximately 300 kilometres northwest of the Burgess Shale.

The Cambrian Explosion

The early Cambrian Period marks one of the most spectacular evolutionary events in the history of life on Earth. Within ten million years (about 520 to 530 million years ago), a very short period geologically, a host of hard-bodied animals appeared in the fossil record. Trilobites, archaeocyathids (cone-shaped creatures possibly related to sponges), molluscs and brachiopods all appeared abruptly. However, as indicated above, this may be an artefact of preservation and the rarity of soft-bodied fossils. There may be no fossils of trilobites in the lowermost Cambrian rocks, but paleontologists know that soft-bodied trilobite ancestors roamed the sea floor at that time because these primitive animals left diagnostic tracks in the mud.

Debate still rages as to the cause of the sudden development of skeletons. Skeleton formation may have been due to changes in atmosphere and ocean chemistry — notably an increase in atmospheric oxygen level — and the development of metabolic pathways allowing biomineralization. In addition, predation may have exerted selection pressure which favoured survival of those with protective hard parts.

The Burgess Fossils — Classifiable or "Weird Wonders"?

Paleontologists describe fossils and classify them based on physical similarities with other known species. Walcott went at this task with a passion, naming and describing over 100 of the 170 species presently known from the Burgess Shale. Walcott's classifications did not receive a thorough re-examination until Harry Whittington and his graduate students Simon Conway Morris and Derek Briggs took up the work in the 1970s.

Harry Whittington began studying a set of arthropods. This diverse group of animals have the common characteristics of having jointed legs and segmented bodies. To classify these creatures paleontologists examine the legs: their structure, number and location. Based on these criteria all previously known arthropods could be classified into four main groupings: uniramians, such as the insects, that lack gill branches on their legs; crustaceans, such as crabs and shrimp, that all have five pairs of appendages on their heads, two in front of the mouth and three behind; chelicerates, including spiders and scorpions, all have six pairs of appendages on the front end of their bodies; and trilobites, an extinct class of arthropod that all had four pairs of head appendages, one pair in front of the mouth and three behind.

Through meticulous work, Whittington realized that many of the animals of the Burgess Shale lay outside of these four groups. First there was *Marrella* with only two pairs of appendages on its head. Next he examined *Yohoia* and discovered that it had only one pair of appendages on its head, and these were unique with their two segments and four clasping fingers on each arm. And *Opabinia*, with its five stalked eyes and long frontal appendage ending in a pincer with grasping spines, was unlike any other animal then known.

Whittington and his students went on to redescribe and reclassify many Burgess Shale animals. *Marrella, Yohoia, Leanchoilia*, and *Burgessia* are considered arthropods, but do not fall into any of the four established classes. Collins recently placed *Anomalocaris* and *Opabinia* within the arthropods. *Amiskwia, Nectocaris*, and *Dinomischus* are not currently assigned to major groups, and the assignment of *Wiwaxia* to the polychaetes is likely to change. Debate continues as to whether the unclassified organisms constitute extinct phyla or simply reflect our ignorance of the full range of early members of the existing phyla.

Perhaps we create our own difficulties by trying to classify these organisms into a too-rigid framework. The traditional system of Linnaean classification relies on *shared* characteristics to group organisms. It is biased against intermediate forms. An alternative system is cladistics, which emphasizes *derived* characteristics. Cladistics recognizes that organisms change over time but members of a natural grouping — a *clade* — share unique features which were not present in a distant common ancestor.

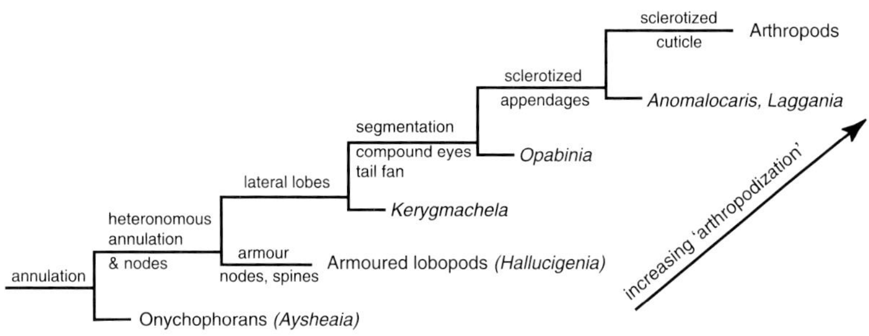

Cladistic classification concentrates on derived characteristics, and shows the links several hitherto "unclassifiable" organisms have to the arthropod phylum. (Simplified from the work of Graham Budd, an Uppsala University paleontologist.)

Charles Doolittle Walcott

Born in 1850, Charles Walcott had no formal training in geology. However, he lived in upstate New York, where the Ordovician bedrock abounds in fossils. There he had the good fortune to cross paths with several laymen who passed on to him an appreciation of fossils, their collection and preparation. By age twenty three, Walcott was sufficiently accomplished that he sold a large fossil collection to the Harvard geologist Louis Agassiz.

Walcott's industriousness and attention to detall attracted the notice of several geologists, and in 1879 he became a temporary geological assistant with the fledgling United States Geological Survey (USGS). His interest in trilobites, honed on the farmlands of New York State, flourished through his assignments on the Cambrian System. By 1894 Walcott had risen to the top of the USGS, becoming its third director, and in 1907 he was appointed secretary (that is, head) of the Smithsonian Institution.

Walcott had become an authority on Cambrian fossils, and it was this interest that drew him to western Canada. He spent every field season from 1907 to 1925 (only two years before his death) in the Rockies, mapping the thick Cambrian formations from Mt. Robson to Banff. The culmination of his career was the discovery of the soft-bodied fossils in the Burgess Shale in 1909.

Walcott shipped his Burgess Shale fossils — 65,000 specimens on 30,000 slabs — back to the Smithsonian Institution in Washington. His duties there prevented him from studying them in much detail, but he was able to describe most of the fauna in five volumes of the Smithsonian Institution Miscellaneous Collections. Many of his notes and photographs lay in the drawers alongside the slabs for nearly half a century, until the modern generation of researchers dusted them off, beginning in 1973.

Life in Camp

Helena Stevens Walcott was Walcott's second wife. Following their marriage in 1888 Walcott, as a true geologist, talked her into honeymooning in Newfoundland so that he could collect Cambrian trilobites. This set the tone of their life together.

Helena gave birth to Sidney, Helen, and Benjamin. With the discovery of the Burgess Shale, all the Walcotts assisted in fossil collecting. These summers may have been the happiest time of Walcott's life. He was camped in a lovely place with dramatic scenery, his wife (photo, right) and three of his four children were with him, and he had his hands on simply incredible fossils. Blocks were quarried out, slid down slope to the trail, and then packed by horse to the camp. Helena did essentially no climbing up to the quarry, but spent day after day in camp splitting the shale blocks and accumulating treasures for her husband.

Arthur Brown worked as a messenger for Walcott in Washington. However, his real strength lay in his ability to organize and cook for Walcott's field camps in the Rocky Mountains.

From 1908 to 1925, Arthur Brown never missed a field season. When the party arrived at a new camp site, they would be tired, hungry, and perhaps chilled. There is no affection like that of a trail crew for a cook who could quickly set hot food before them.

In all of Walcott's diaries there is never a note of Arthur being ill, though the injuries and sickness of others are recorded. Day in and day out, year after year, Arthur Brown kept the camp running smoothly. At age 75 Walcott had his final field season. He would never have been able to accomplish as much as he did in the Canadian Rockies without the support of Arthur Brown.

—Ellis Yochelson

The
Burgess
Shale
Quarries

The Walcott Quarry is the most prolific of the fossil quarries on the western slope of Fossil Ridge. First excavated by Charles Walcott between 1910 and 1924, it was extended to the north by the Geological Survey of Canada in 1966–67 and then dramatically enlarged by the Royal Ontario Museum starting in 1993 under the leadership of Desmond Collins.

The fossils are unevenly distributed. Most layers are barren; others are unbelievably rich. The best soft-bodied fossils occur in a section 2.3 metres thick that Walcott called the Phyllopod bed. (*Phyllopod* is an old term for a group of crustaceans with rows of leaf-like gills running down their sides.) The Phyllopod bed is now part of an informal rock unit, seven metres thick, called the "Greater Phyllopod Bed." The Royal Ontario Museum collected over 100,000 fossils from the Greater Phyllopod Bed between 1993 and 2000. The Museum has become the world's largest repository of Burgess Shale fossils, surpassing the Walcott collection at the Smithsonian Institution in quality and quantity.

Additional fossil beds exist at levels above and below the Walcott Quarry. The Raymond Quarry, named after a Harvard University paleontologist who excavated in 1930, lies about 20 metres above the Walcott Quarry. It has yielded some good examples of sponges and worms, but the overall quality of preservation is lower.

Over the two million years that it took to fill the basin in front of the Cathedral Escarpment, mud settled through the water slowly but continuously, gradually building up layers of sediment on the sea floor. These layers are usually barren of fossils or animal tracks and burrows, indicating the environment was hostile to life — perhaps low in oxygen. Reconstructions based on fossil zones and environmental indicators suggest the water depth at the foot of the Cathedral Escarpment was about 100 metres. This depth is sufficient to provide a quiet water

Alternating dark and light banding marks different shale layers in the Walcott Quarry. Coin for scale measures 2.5 centimetres. Fossils with soft body preservation tend to be in the darker siliceous mudstone bands.

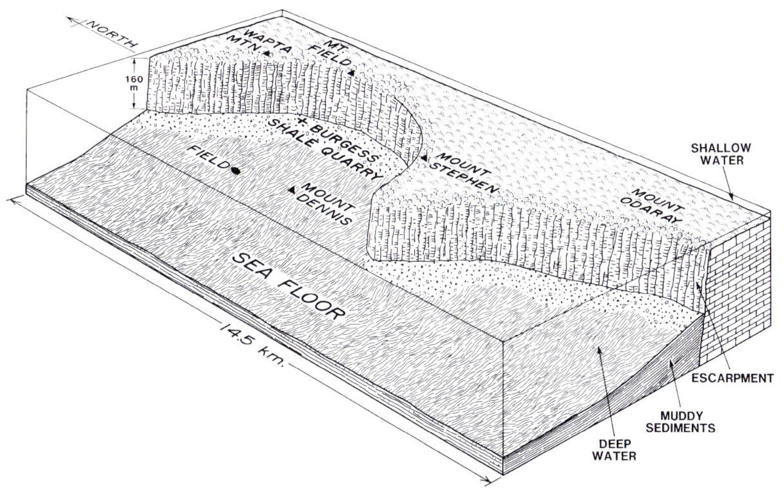

Above: Reconstruction of the Cathedral Escarpment, with present-day landmarks superimposed.

Below: Artist's reconstruction of a mudslide burying the Burgess Shale fauna.

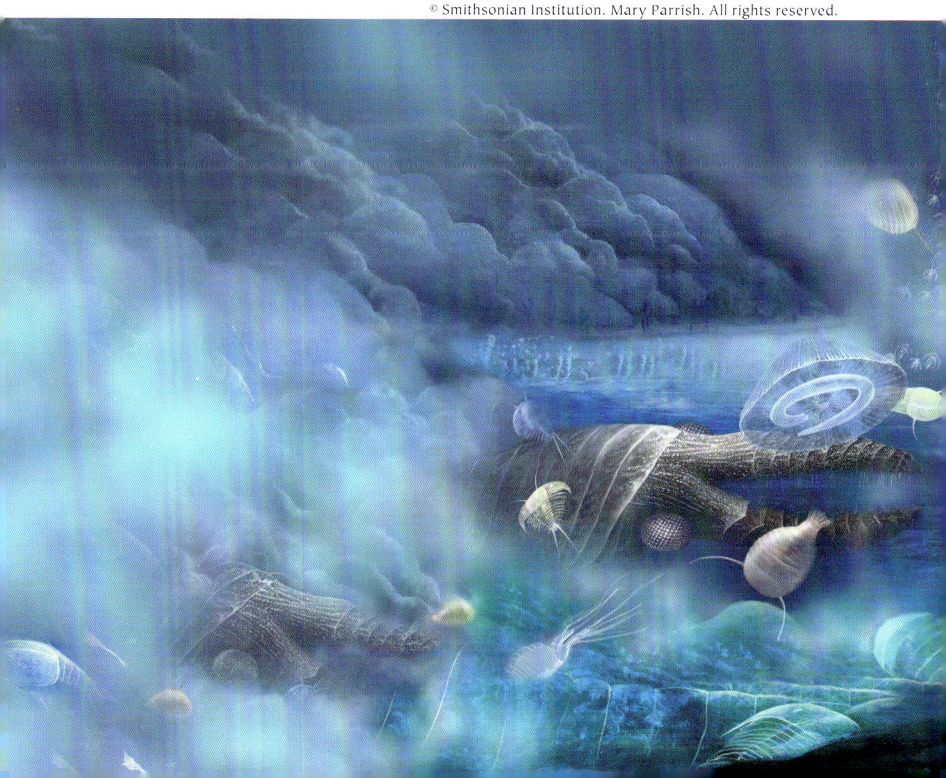

environment but shallow enough to permit some light to penetrate.

The source of the muds lay to the present-day north. Wind-driven oceanic currents moved the muds parallel to the front of the escarpment, where they were deposited as a wedge that thins from the base of the escarpment out into the basin. Periodically, these muds slumped downslope as a result of instabilities created by loading or triggering by earthquakes. When this happened, any creatures living on the sea floor were swept along, suffocated and deposited with the turbid mud flows.

Walcott failed to record the exact stratigraphic position of his specimens, thus losing the fine-scale temporal record of the community. Fossils collected from the Greater Phyllopod Bed by the Royal Ontario Museum were collected bed by bed, allowing for study of the community in its proper stratigraphic context. This provided better understanding of how the community was buried and how it changed through time.

Dr. Jean-Bernard Caron from the Royal Ontario Museum studied dozens of bedding assemblages representing over 60,000 fossil specimens. He showed that the Greater Phyllopod Bed community was mostly represented by animals living at the sea surface or near the sea floor. Pelagic animals (swimmers living in the water column) like the giant predator *Anomalocaris* were extremely rare. He concluded that most of the fossilized animals were buried very quickly and close to where they lived by sudden mudflow events. The Raymond Quarry strata have more burrows and algae suggesting that their environment was shallower and more oxygenated than that of the Greater Phyllopod Bed. Quantitative comparisons of different levels indicated that the number of species and individuals did not vary much through time but that a turnover pattern existed wherein many species, especially the rare ones, tended to be replaced by other species in younger levels.

Fossil Ridge, seen from Burgess Pass, runs between Wapta Mountain (left) and Mt. Field (right). The quarries lie between treeline and the shadow, at the arrow position, centre of photo.

Soft-body Preservation

Almost all fossil beds in the geologic record contain only the remains of animal skeletons: shells, bones, teeth. However, animals with mineralized "shelly" parts account for only one quarter of the animals living in Earth's oceans today. Paleontologists assume that most animals are not normally preserved in the fossil record. Reconstructing and understanding ancient animal communities is like trying to determine the picture in a jigsaw puzzle when three quarters of the pieces are missing. Rare preservation of soft-bodied fossils, such as in the Burgess Shale, provides glimpses into these lost worlds.

What unusual conditions combined to allow for the spectacular fossil preservation in the Burgess Shale? Most researchers agree on two absolute requirements: the carcass must be hidden from scavengers and buried quickly in an anoxic environment. The mudflows achieved this by burying the carcasses in an oxygen-starved environment where they could not be reached by Cambrian scavengers. But there is more to it than a simple lack of oxygen, as many bacteria can decompose an animal without it.

Some paleontologists believe that the answer to the preservation puzzle lies in the rock surrounding the fossils. The mudstones consist almost entirely of clay. One remarkable characteristic of clay is that many organic molecules, including those necessary for decomposition, will stick to the surfaces of the tiny grains (less than four thousandths of a millimetre in diameter). However other authors have disagreed with this hypothesis. The chemical conditions in which soft tissues are preserved are still poorly known and hotly debated today.

Most of the Burgess Shale fossils were laid down flat, indicating a gentle form of burial. However, they have clay intercalations separating each delicate appendage. The clay inhibited decomposition, leaving the thin films that define the Burgess Shale fossils today.

Part and Counterpart

Fossils are revealed when the enclosing rock is split, separating the fossil into a part and counterpart. Walcott worked hard at photographing the two-dimensional silvery films on the part and counterpart, sometimes heavily retouching his photos to bring out low-contrast details. Harry Whittington realized that the split at the fossil tends to occur in microscopic stair steps linking the clay partings that are interlayered between the body parts. The part and counterpart are not identical because different body parts are concealed beneath each surface. Modern preparators use modified dental drills to carefully remove the clay interlayers to expose unseen details that differ between the part and counterpart. This method is responsible for the new understanding of the Burgess fauna initiated by the Cambridge project.

Quarrying Method

The Royal Ontario Museum research team relied on a lot of muscle and a little engineering to excavate the Burgess Shale. Along the back wall of the quarry you can see sets of vertical drill holes that were used to break the shale without explosives. One worker drilled a line of holes into which steel wedges were inserted. Next a team of workers used sledgehammers to drive the wedges deeper, eventually creating a vertical fracture separating a block of rock from the back wall of the quarry. Using crowbars, each block was tipped onto its side for coarse splitting. Not all layers have fossils and collectors, after some experience, learned which layers had to be finely split using a much thinner chisel. When a fossil was found, part and counterpart were collected and its stratigraphic position recorded. If the layer was barren, then it was moved out of the way, down the mountainside.

Almost all of the talus below the quarries was split and discarded in the search for fossils.

Protection from Tectonism

The well-bedded, black shales of the Burgess Shale Formation end abruptly at the north end of the quarry, beside a rubble filled gully that runs up Fossil Ridge. Across the gully, the sparse rock outcrop is rusty brown and lacks bedding. These rocks are dolomitized limestones of the Cathedral Formation. The gully marks the vertical contact between the Cathedral and Burgess Shale Formations, and therefore, the Cathedral Escarpment. Proximity to the butressing escarpment is what may have saved the Burgess fossils from destruction in subsequent rock-shearing episodes of compaction under ten kilometres of younger sediment and subsequent exhumation and mountain building.

The Evolution of Walcott's Quarry

Charles Walcott watches his son, Stuart, at work in the Burgess Shale quarry in the summer of 1913. W.J. Stevenson and R.D. Messler work on the quarry face. Large slabs, propped in the foreground, await splitting or trimming. The quarry size was relatively modest, limited by manpower and the short field season. When Walcott's field work ended, just over a decade later, he considered the fossil bed more or less exhausted of specimens.

After the Geological Survey of Canada collected in the late 1960s, nature took over and the quarry began to infill with rock debris from frost action and talus slides.

The Royal Ontario Museum has mounted major research and collecting expeditions to the Burgess Shale in the last 25 years. This 1999 view shows the massive amounts of rock which have been moved in expanding the quarry. Led by Dr. Desmond Collins, researchers have uncovered many fine specimens — some new to science — and shed new light on the paleoecology of the Cambrian sea floor.

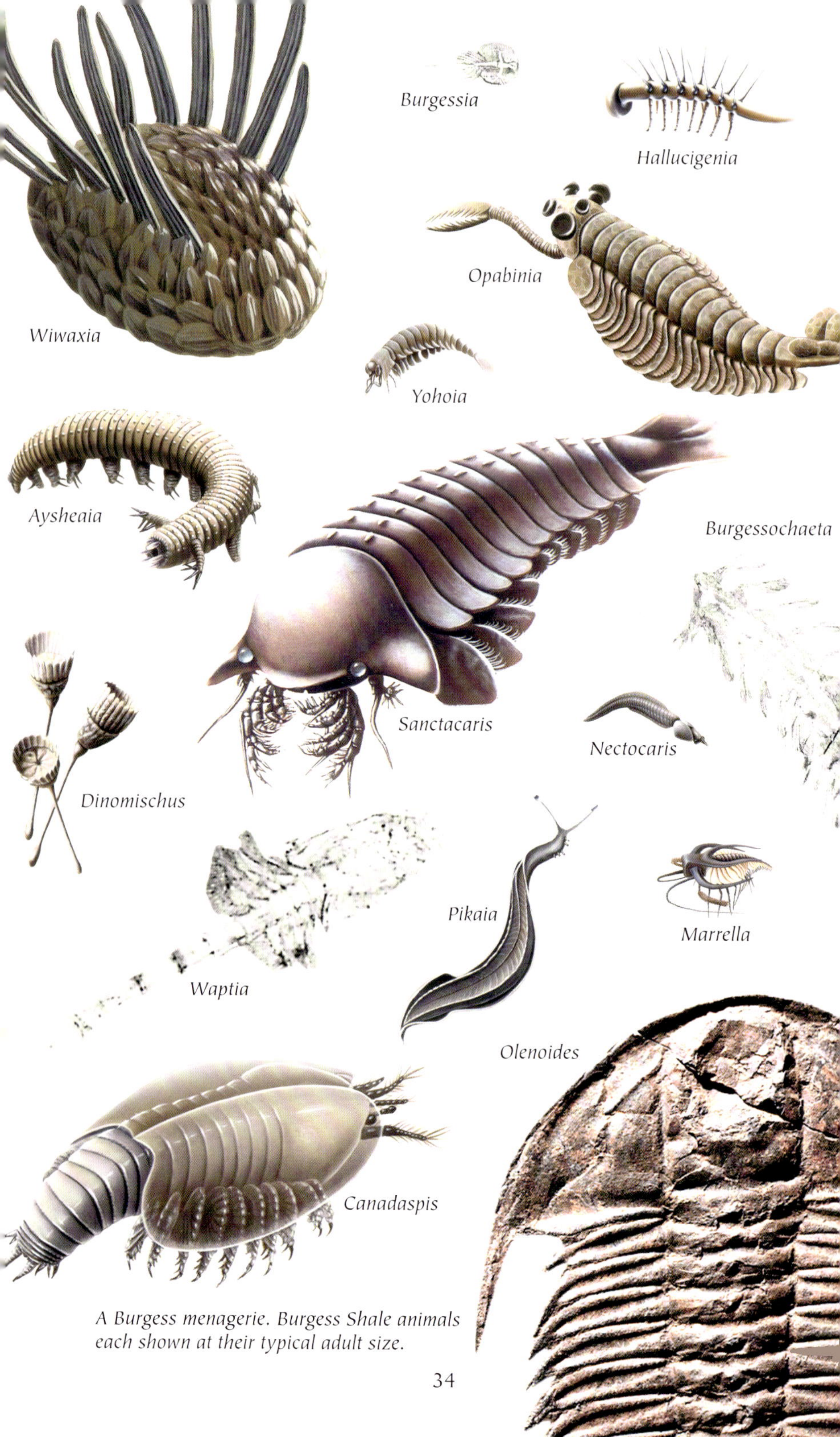

A Burgess menagerie. Burgess Shale animals each shown at their typical adult size.

Fossils of the Burgess Shale

Aysheaia (Phylum Onychophora)
Aysheaia is a lobopod (a primitive sister group to the arthropods), so-named because of its lobe-like walking limbs which support an annulated trunk. It has a forward-facing mouth surrounded by papillae that aided feeding, a pair of spiny appendages near the head, and the limbs have small claws. It is often found associated with sponges, and may have used its appendages to hold and feed on them.

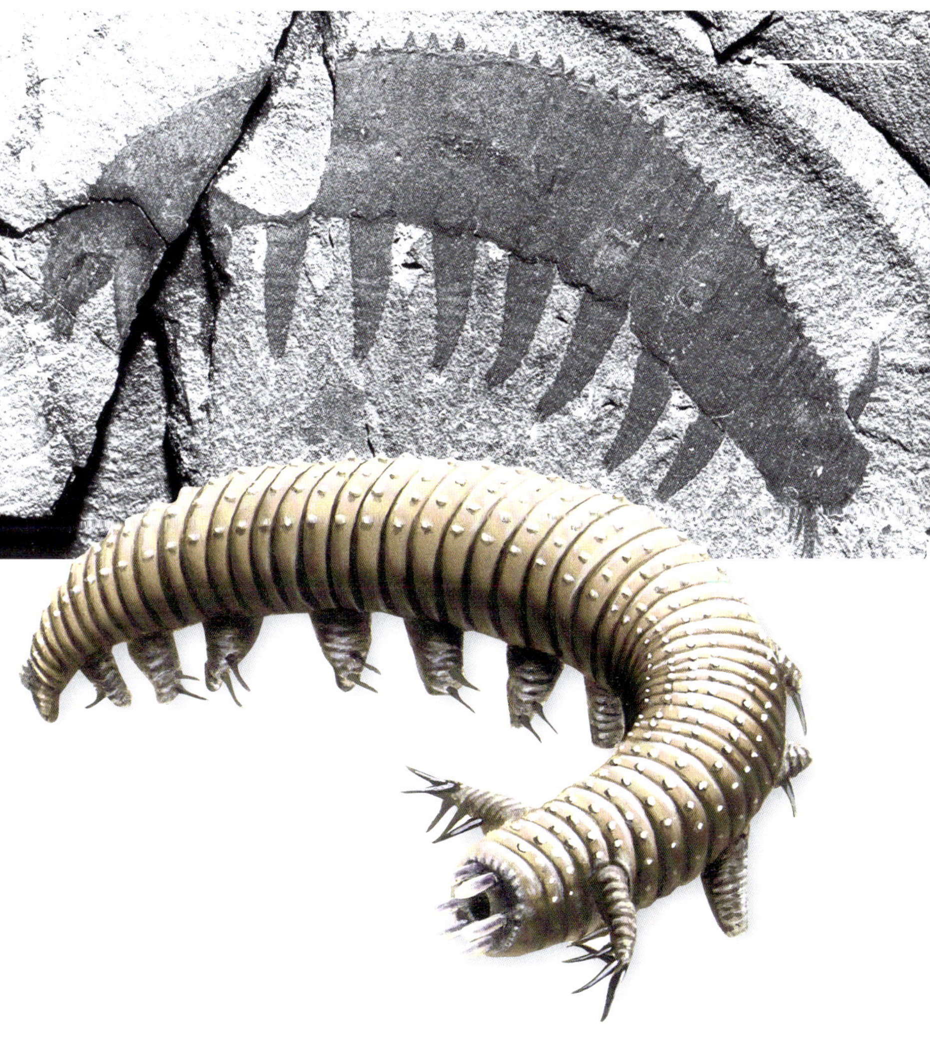

Hallucigenia (Phylum Onychophora)
Burgess Shale specimens of *Hallucigenia* are difficult to interpret. The original reconstruction by Simon Conway Morris showed *Hallucigenia* walking on pairs of pointed stilt-like legs, with a single row of feeding tentacles projecting from its back. His consternation at this arrangement resulted in its name. Today, with the benefit of better-preserved specimens found in China and careful dissection of a Burgess specimen, which revealed a second row of tentacles, each with a tiny claw, the reconstruction has been inverted and *Hallucigenia* is interpreted as an armoured lobopod. The form of the anterior end (the head) of *Hallucigenia* is poorly known. Interpretations range from a mushroom cap shape to a mere swelling of the trunk.

Marrella (Phylum Arthropoda, Primitive)

Marrella is the most common fossil of the Burgess Shale, with over 25,000 specimens collected. Two pairs of spines curving back from the head shield probably served as light armour. Delicate filaments along its gill branches are usually well preserved and were responsible for Walcott calling *Marrella* the "lace crab". *Marrella* fossils often show a dark stain extending out from the tail. This is the remains of fluids that leaked from the body soon after burial. *Marrella* likely walked the sea floor sweeping food into its mouth with its brush-like head appendages.

Burgessochaeta (Phylum Annelida, Class Polychaeta)

Burgessochaeta is a polychaete (bristle worm) with a trunk organized into at least twenty four segments. Each segment bears four bundles of bristles (chaetae), two on either side of the body. Each bundle consists of 11 to 17 chaetae which the animal used to propel itself. A pair of anterior tentacles (not visible in photo) were used as sensory organs and probably aided feeding.

Canadaspis (Phylum Arthropoda)
As a crustacean, *Canadaspis* is an ancient relative of modern shrimp. Its body was protected by a hinged carapace. Over 4,000 specimens are known in the Walcott Burgess Shale collection, most preserving only the carapace. The limbs of *Canadaspis* were biramous, consisting of a walking leg and a gill branch. *Canadaspis* swam and fed on small animals and material that it churned up from the mud with its legs.

Waptia (Phylum Arthropoda)
Waptia also covered its body with a carapace. A long, legless abdomen extended far beyond the shell, ending in a forked tail that looks much like that of a fish turned on its side. Sensory organs in front included two stalked eyes and a pair of antennae. Unlike *Canadaspis*, *Waptia*'s limbs were uniramous, with separate limbs for walking and breathing. *Waptia* both swam and walked on the sea floor, where it foraged for organic particles in the mud.

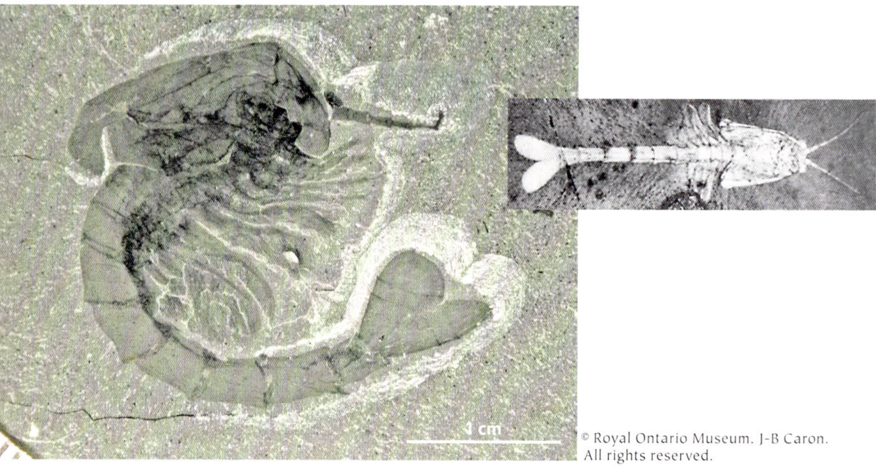

Burgessia (Phylum Arthropoda, Class not assigned)
Although less than two centimetres long, this fossil is often spectacular because of the preservation of its delicate, branching gut that can often be seen below its notched, circular carapace. *Burgessia* was a blind animal that foraged along the sea floor, probing with a pair of front antennae. Although it is a unique class of arthropod, it falls within the larger group of arthropods known as arachnomorphs, which includes the trilobites and chelicerates (e.g., spiders).

Sidneyia (Phylum Arthropoda, Class not assigned)
Sidneyia was a large predator (up to 13 centimetres) that stalked its prey across the sea floor. Its diet included hyolithids, ostracods and trilobites, all of which have been found preserved in its gut. *Sidneyia* had a short, wide thorax, blunt head with two antennae, and a tail ending in a broad fan. This animal bears the name of its discoverer, Sidney, the second son of Charles Walcott.

Leanchoilia (Phylum Arthropoda, Class not assigned)
What clearly sets *Leanchoilia* apart from most other arthropods are its two front appendages that each branch into three whip-like arms. *Leanchoilia* lacks the spiny limbs characteristic of predators, and this blind animal probably used its whips as sensory organs as it scavenged for food in the mud. The railway siding of Leanchoil, near the present Chancellor Peak campground in Yoho National Park, lent its name to this ancient creature.

Sanctacaris (Phylum Arthropoda, Class uncertain)
Sanctacaris was not discovered until 1981. It had a wide head shield beneath which six pairs of feeding appendages emerged. The streamlined head and wide flat tail suggest *Sanctacaris* was a strong swimmer. It was once classed as a chelicerate, a forerunner of today's horseshoe crab. However it is currently placed in the Arachnomorpha, a broader ancestral group of assorted basal forms that gave rise to the trilobites and chelicerates.

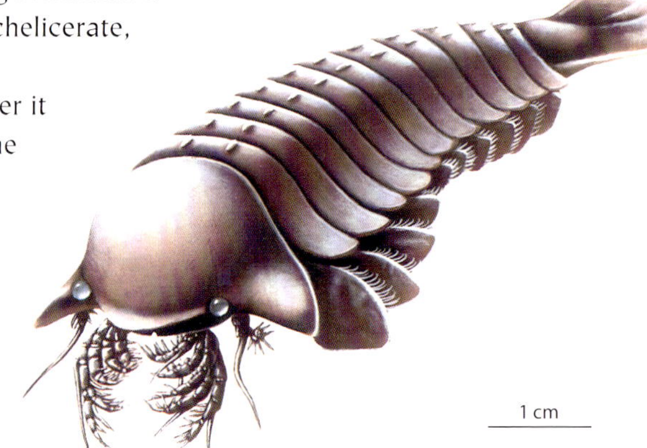

Yohoia (Phylum Arthropoda, Class not assigned)

Yohoia was a sleek creature, up to 2.5 centimetres in length, which possessed two articulated front appendages each with four terminal spines. Its form has been compared to the modern mantis shrimp, which is much larger (up to 60 cm) and uses its front appendages to spear or smash shelled organisms in order to get at the soft tissue. The small size of *Yohoia* suggests it was a relatively delicate creature that probed for food particles on the seafloor and might have been able to prey on larval trilobites or other small soft organisms.

Wiwaxia (Phylum not assigned)

Wiwaxia is a slug-like creature whose top surface was covered with leaf-shape ribbed plates (sclerites) and two rows of longer spines. These are often preserved as a flattened mass of armour, as in the illustration at right, which hides the details of the soft tissue. Occasionally a radula bearing two rows of teeth is seen at the anterior (head) end of the organism. *Wiwaxia* has been considered a polychaete (bristle worm), but this interpretation is controversial. It crawled along the sea floor, feeding on organic detritus.

Nectocaris (Phylum not assigned)

Nectocaris is extremely rare in the Burgess Shale which, together with its streamlined body, suggests it was a swimmer unlikely to have been caught in mudflows. Its head is protected by a pair of oval shields. With large eyes and a pair of frontal appendages, *Nectocaris* was probably a swift-moving predator.

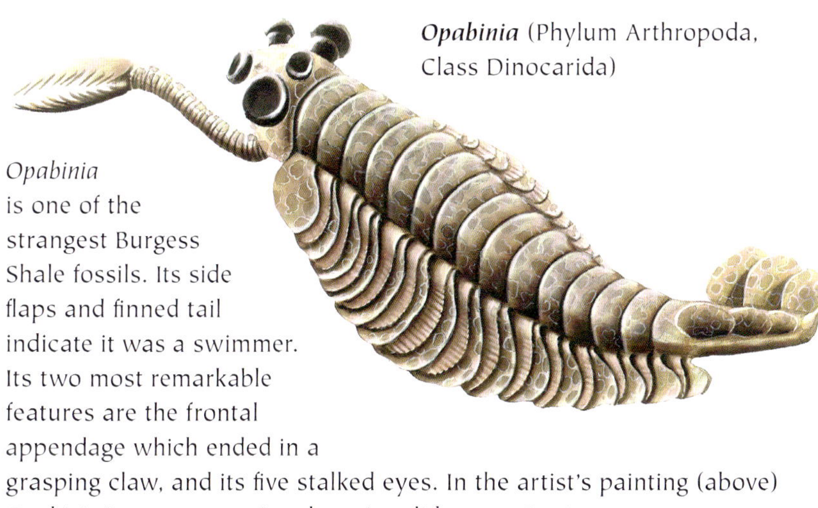

Opabinia (Phylum Arthropoda, Class Dinocarida)

Opabinia is one of the strangest Burgess Shale fossils. Its side flaps and finned tail indicate it was a swimmer. Its two most remarkable features are the frontal appendage which ended in a grasping claw, and its five stalked eyes. In the artist's painting (above) *Opabinia* is seen capturing the priapulid worm *Ottoia*.

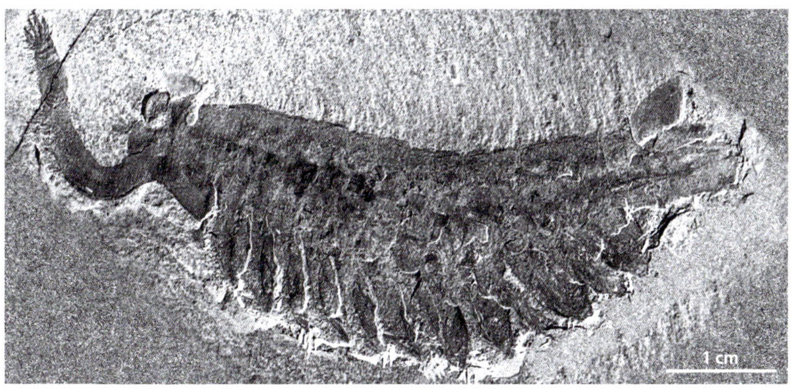

Anomalocaris (Phylum Arthropoda, Class Dinocarida)
Fossils that resemble headless shrimp (lower left) were first described by J.F. Whiteaves in 1892, and this label was latinized to become *Anomalocaris*. Walcott collected circular fossils that he thought were a kind of jellyfish and christened *Peytoia* (lower right). Decades later, Harry Whittington and Derek Briggs discovered that these fossils were parts of a single organism. The "shrimp" was one of a pair of feeding appendages, and the "jellyfish" was the mouth of the largest predator in the Cambrian seas. *Anomalocaris* was up to half a metre long, swam by means of rhythmic undulations of its side lobes, and fed its prey into its circular mouth with its two spiked feeding arms. Recent redescription by Desmond Collins places *Anomalocaris* and its sister species *Laggania* in an extinct class of arthropods, the Dinocarida (*terrible crab*). *Anomalocaris* possessed stalked eyes and a tail comprised of three paired fins. Trilobites have been found with healed injuries that could have been caused by *Anomalocaris* bites.

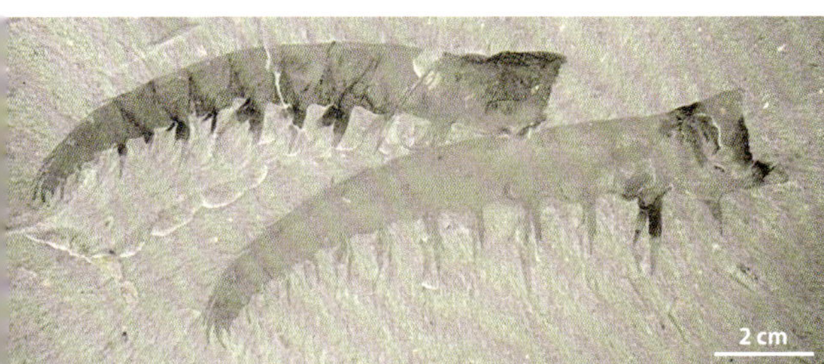

Above: Disarticulated claws (left) and mouth parts (right) were initially thought to be separate organisms. Below: Anomalocarids on the hunt.

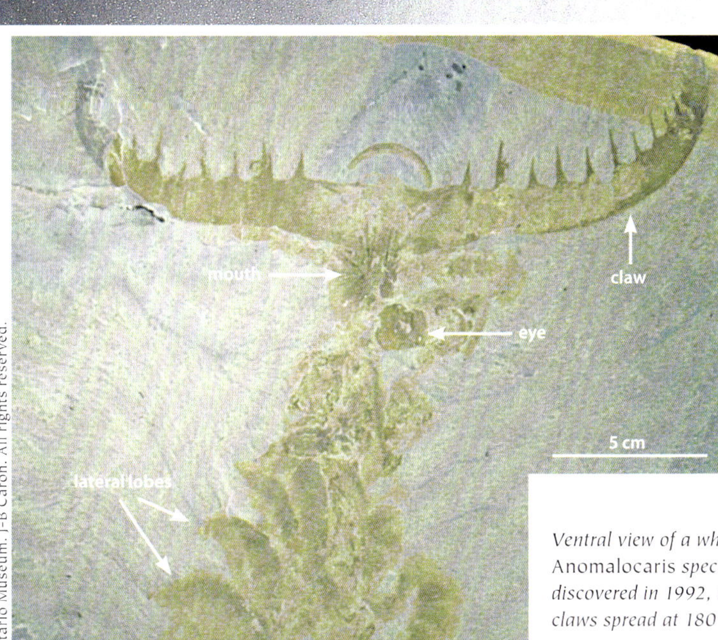

Ventral view of a whole Anomalocaris *specimen discovered in 1992, with claws spread at 180°.*

Naraoia (Phylum Arthropoda, Class not assigned)
Naraoia is characterized by anterior and posterior ovoid shields covering its body, and biramous walking legs and gill branches. The dorsal view (left photo) shows faint traces of gut diverticulae in the anterior segment, and a portion of the right antenna (arrow). The walking legs and gill branches are evident in the lateral view (right photo). Both images show axial soft tissue preserved as dark mineralized material. *Naraoia* was long considered to be a primitive soft-bodied trilobite, but recent work indicates naraoiids are not trilobites, nor are they closely related.

Dinomischus (Phylum not assigned)
Dinomischus was a small animal with a flexible bulbous stem that anchored it to the sea floor. It was a filter-feeder, with the gut contained in the main body, or calyx, at the top of the stem. A fringe of bracts probably helped to direct water currents to the central mouth. [x2]

Pikaia (Phylum Chordata)

Pikaia is considered to be a chordate, but this animal still awaits a proper description. It swam like a modern fish, by means of muscle contractions undulating along its body and working against a flexible stiffening rod, or *notochord* — a precursor to the vertebrate backbone. *Pikaia* was long thought to be an early ancestor of the whole vertebrate lineage, including man. However, it possesses some uncharacteristic features, including a head consisting of two lobes, each bearing an appendage. The 1999 discovery of primitive fish in the Lower Cambrian of China suggests both they, and *Pikaia*, are descendants of an even-older ancestor.

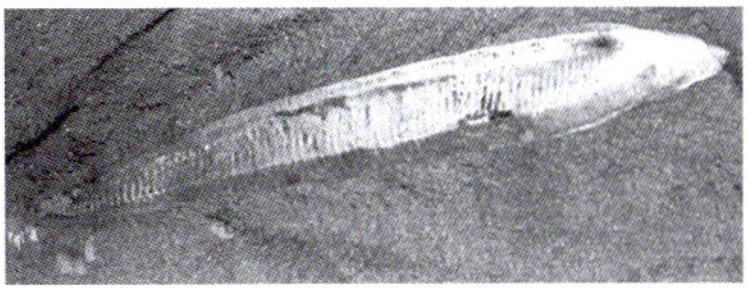

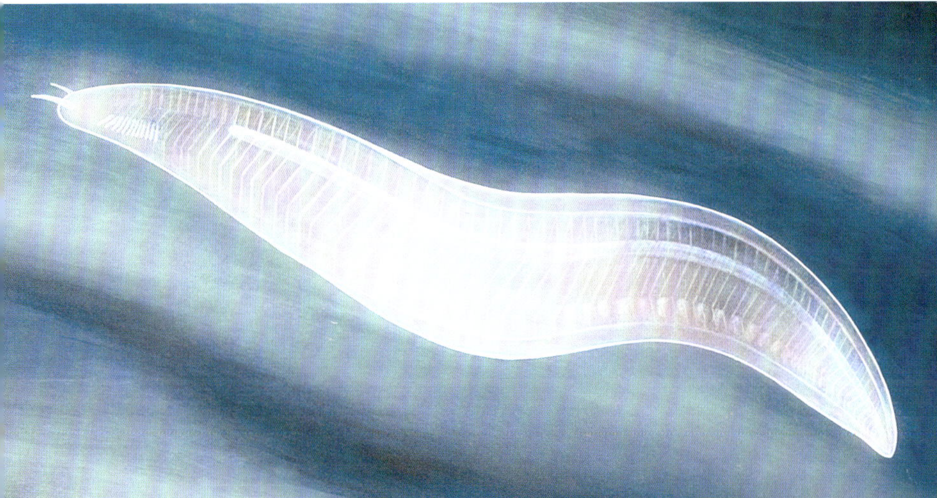

Two Views of Evolution

Harvard paleontologist Stephen Jay Gould was struck by the variety and unusual features of many of the Burgess fossils and the seeming difficulty of classifying them into one or another of today's phyla. In his book *Wonderful Life*, a popular account of the Burgess Shale, he put forward the idea that the Cambrian explosion allowed a multitude of animal designs to proliferate, but that many were subsequently lost through the rise of competition and a series of chance extinctions.

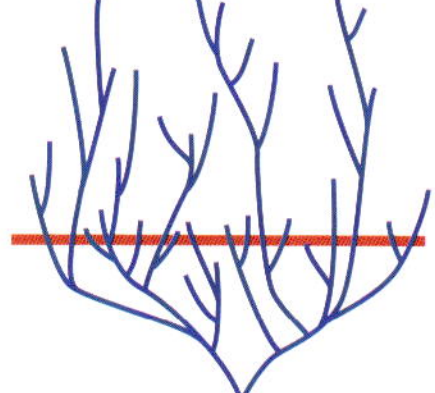

Gould's bush (left) has maximum disparity in the Cambrian (red line). Conway Morris favours a rapid disparity increase in the Cambrian, but continuing to increase irregularly through time (right).

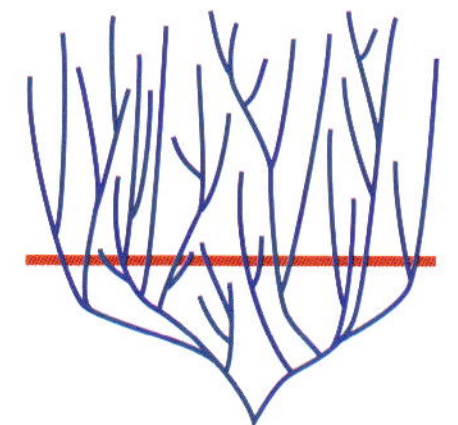

Gould argued against the concept of an upward-radiating Tree of Life suggesting instead that a bush, with many initial branches but only a few persisting, is a better metaphor. His most famous assertion was humans are but one possible outcome of speciation and extinction, and if the "tape of life" was rewound and played again a completely different result could emerge.

Simon Conway Morris, is a Cambridge paleontologist who has worked on the Burgess Shale for over thirty years. He published a different view in *Crucible of Creation*. While acknowledging the variety of Burgess life forms, Conway Morris believes that basic underlying features confirm their placement within known phyla. His view is that the body designs we see on Earth are predetermined by the environments available. A large animal that swims requires a streamlined body, fins for directional control, and a strong backbone and muscles. Thus icthyosaurs (reptiles) and whales (mammals) converged on these body features. The Cambrian explosion was a time when disparity (design forms) increased rapidly. Thereafter it increased at variable rates. In Conway Morris' view, re-running the tape of life would produce more or less the same results, different only in detail.

Trilobites

Trilobites come close to being everyone's favourite fossil. Trilobite fossils are easy to comprehend. Most range in size from one to ten centimetres, small enough that the entire body can be preserved on a fist-size rock slab, yet large enough to study with the naked eye or a hand lens.

Trilobites are marine animals belonging to a phylum which includes insects, centipedes and millipedes, crustaceans and spiders. The name *arthropod* comes from the Greek words for *joint* and *foot*, appendages which are the defining features of this phylum. This body plan has been extremely successful — approximately three-fourths of all known animals are arthropods!

Trilobites appear in the fossil record about 15 million years after the dawn of the Cambrian Period. They were present on Earth for nearly 300 million years before becoming extinct at the end of the Permian. During this time they underwent several evolutionary events.

At the end of the Cambrian, many unspecialized types became extinct. Diverse new forms radiated out into new environments during the Ordovician. A major extinction 367 million years ago, near the end of the Devonian, wiped out most trilobites. Only one trilobite order survived into the Permian. Several thousand trilobite species are known. This diversity helps paleontologists date and correlate rocks from different areas based on the particular trilobite assemblages they contain.

Trilobites are so-named because their body plan consists of three lengthwise lobes — a central **axial lobe** flanked by two **pleural lobes**. They had an external skeleton, or **exoskeleton**, formed of microcrystalline calcite needles set in an organic matrix. The exoskeleton provided protection and a firm base for muscle attachment. Articulating segments allowed for movement and enabled the trilobite to roll up into a ball to protect its vulnerable underside.

The head end of a trilobite is called the **cephalon**. The **glabella** is a bulbous central area on the cephalon, which housed the stomach. On the underside is the **hypostome**, a plate which covered the mouth. The glabella is flanked by the **fixed cheeks** and **eye ridges** (in trilobites that had vision). **Genal spines** of varying length, where present, served to spread the weight of the trilobite on the soft sea floor and to act as defensive projections on an enrolled body.

The central part of the trilobite is called the **thorax**. It was composed of articulated segments consisting of an **axial ring** and two **pleurae**. Unmineralized linkages allowed the body to flex. The posterior end, the **pygidium**, is comprised of fused segments.

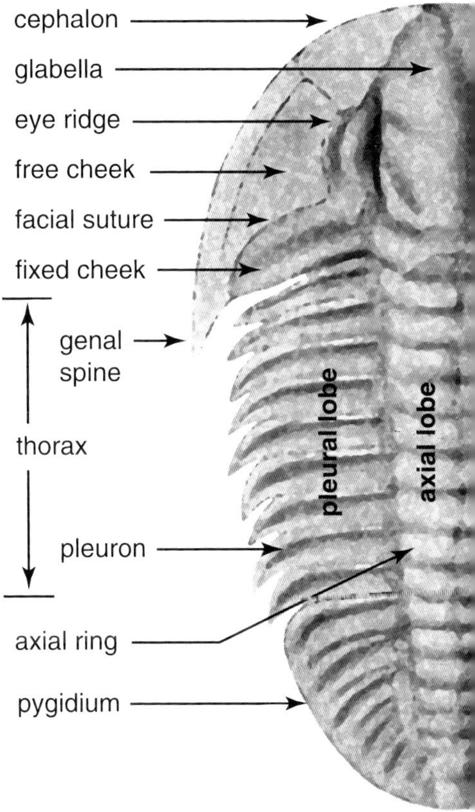

There was one disadvantage to an exoskeleton: trilobites had to periodically shed and secrete a new, larger one in order to grow. Most did this by shedding their **free cheeks** along a **facial suture**, and wriggling out of the front of the exoskeleton. If you find a trilobite fossil with a complete semicircular head, then it is the remains of a dead animal. If the head has large indentations on both sides, then the free cheeks are missing and that fossil is probably a moulted exoskeleton.

Trilobite Lifestyles

Olenoides serratus is probably the world's best known trilobite, due to the preservation of appendages in about one hundred specimens from the Burgess Shale Formation. The specimen below shows several appendages protruding (lower left) from below the flattened exoskeleton.

The trilobite stomach was located under the cephalon, or head shield. Food entered the stomach through a backward-opening mouth beneath the hypostome, a plate located on the underside of the cephalon, and was directed there by movements of the trilobite limbs.

Paleontologists use the shape and interpreted strength of the hypostome to infer feeding habits of different trilobites. Those with spiny coxae and a hypostome rigidly attached to the cephalon, as in *Olenoides*, probably were predators. They could shred coarse food particles and press them strongly against the mouth. Trilobites with a hypostome attached only by soft tissue fed on smaller particles scavenged from the sea floor or filtered from suspensions created by stirring up mud.

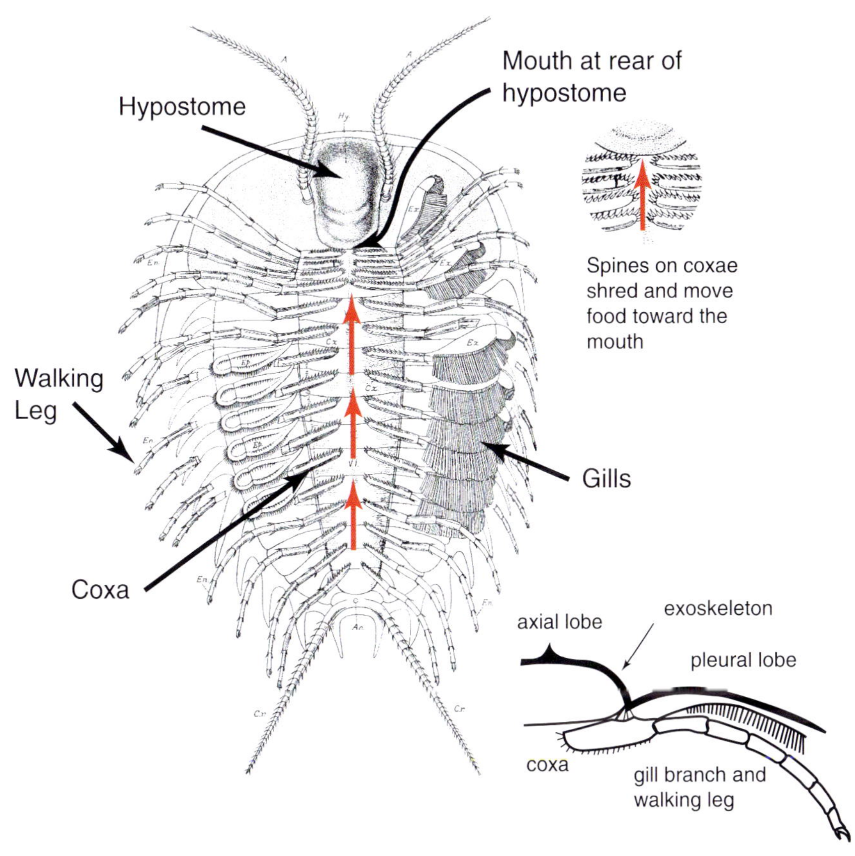

The underside of the Burgess Shale trilobite Olenoides, *as drawn by Charles Walcott in 1921 (main diagram). Right side shows gill branches; left side shows walking leg structure with gill branches removed for clarity. The key to trilobite feeding lies in the spiny coxae, which are capable of shredding soft food and directing it to the mouth (red arrows) at the rear of the hypostome by their reciprocating action.*
A cross section (lower right) shows the relationship of the gill branches, walking legs and coxae.

Trilobites of the Burgess Shale Formation

The Mt. Stephen trilobite beds contain several different trilobite orders. How did so many trilobites manage to eke out a living in the same place? Part of the answer lies in how they partitioned their food resources by adopting different feeding habits.

Mt. Stephen towers over the town of Field. The trilobite beds lie on its western flank, just right of centre.

Agnostida — The Pelagic Feeders

The agnostid genera *Pagetia* and *Ptychagnostus* are small in size with tiny hypostomes. Most are interpreted as swimmers which fed from microscopic plankton and detritus suspended in the water column.

Ptychopariida — The Particle Feeders

Chancia, Ehmaniella and *Elrathina* are ptychopariid trilobites. They are of medium size, with small hypostomes attached to the soft underside tissue. These genera fed on small particles as they scurried across the sea floor or stirred up mud, making shallow depressions. They achieved some protection from predators by living in oxygen-poor, nutrient-rich environments.

Corynexochida — The Predators

The Corynexochida *Zacanthoides, Ogygopsis, Oryctocephalus, Bathyuriscus, Olenoides* and *Kootenia* were predators or scavengers. They are of medium to large size, have rigidly attached hypostomes to cope with the stresses of bigger food particles, and exhibit a variety of spines which were probably useful in defence.

Franco Rasetti—From Physicist to Paleontologist

In 1919, Franco Dino Rasetti was a student at the University of Pisa, along with his friend Enrico Fermi. Rasetti began his professional career as a lecturer in physics at Florence and became part of the small group which, during the 1920s and early 1930s, made Rome a centre of atomic physics. With the approach of the Second World War the group broke up, several physicists coming to the United States. In 1939 Rasetti emigrated to Canada, and in 1940 he began teaching physics at Laval University in Quebec.

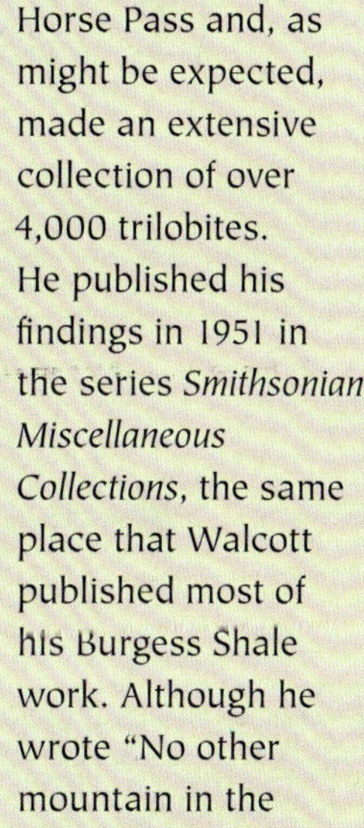

During his years in Italy, Rasetti and his colleagues spent their vacations mountain climbing, and Rasetti was always interested in biology. Indeed, Laura Fermi described him as "a born naturalist" who knew the names of 15,000 insects and nearly as many plants. He became a self-trained paleontologist who taught physics for a living and pursued trilobites as an avocation. He studied trilobites with such diligence and brilliance that in 1952 the United States National Academy of Sciences awarded him the Walcott medal.

Before leaving Canada for the US, Rasetti spent the field seasons of 1947 and 1948 working on the Cambrian of western Canada. He measured sections at fifteen localities centred on the Kicking Horse Pass and, as might be expected, made an extensive collection of over 4,000 trilobites. He published his findings in 1951 in the series *Smithsonian Miscellaneous Collections*, the same place that Walcott published most of his Burgess Shale work. Although he wrote "No other mountain in the area presents stratigraphic and paleontologic problems comparable to those encountered on Mount Stephen," Rasetti was able to use his keen eye for trilobites and his mountaineering skills to sort out the correlations with a precision that still stands today. Franco Rasetti died in 2001, age 100.

—Ellis Yochelson

Pagetia is an agnostid trilobite that fed on small organic particles. Paleontologists are unsure whether it swam or was a bottom dweller. At a maximum size less than one centimetre, the mature form is much smaller than most other fully grown trilobites in the Burgess Shale. *Pagetia* is characterized by having eyes, only two thoracic segments, an annulated pygidium and a terminal axial spine. [x2]

Elrathina has a long, tapering thorax ending in a very small pygidium. In adult specimens the number of thorax segments is variable within a narrow limit — 17, 18 (most common) or 19. This is an unusual feature in mature trilobites. The glabella is broad. [x2.5]

Ptychagnostus lacks eyes on the upper surface of the head shield and probably made its living in the water column rather than on the sea floor. At a maximum mature length of about 10 millimetres, it is much smaller than most other adult trilobites in the Burgess Shale. The glabella has two lobes, the pygidium, three. [x7]

Chancia has a normal glabella but the cephalon is enlarged by an elongate 'pre-glabellar field' ahead of the glabella. Eye ridges are well developed. It possesses genal spines of medium length. The pygidium is very small. [x1.75]

Olenoides (formerly called *Neolenus*) has a large parallel-sided glabella, deep interpleural furrows on the pygidium and slender pygidial spines. *Olenoides* was probably a scavenger as well as a predator. Although relatively few specimens have unmineralized appendages preserved, *Olenoides* is the most common limb-bearing trilobite species in the Burgess/Stephen fauna. [x1.3]

Ehmaniella has a wide cranidium and heavy eye ridges. The pre-glabellar area often shows longitudinal striae. The pygidium is small, with few segments. [x2.5]

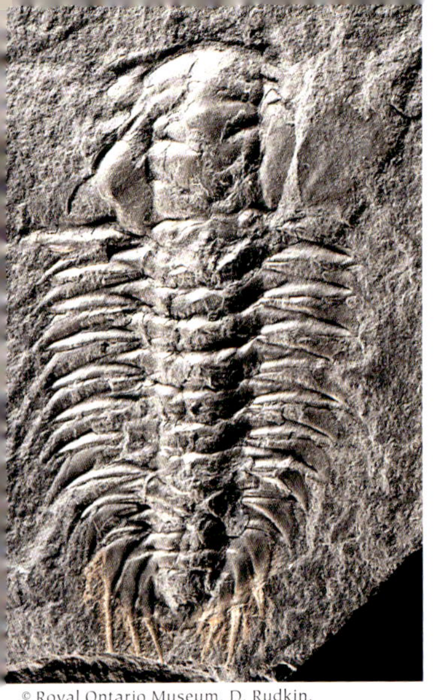

Zacanthoides has a slender exoskeleton with nine thoracic segments. The pleurae have long spines and there are additional spines on the axial rings. Its pygidium is considerably smaller than the cephalon. [x1.5]

Ogygopsis is the most common fossil in the Mt. Stephen fossil beds, but is surprisingly rare in Cambrian faunas elsewhere. It has a prominent glabella with eye ridges. Pleural spines are absent. The large, spineless pygidium of *Ogygopsis* has many segments and is roughly the same length as either its thorax or cephalon. *Ogygopsis* can be up to 12 centimetres in length. [x0.3]

Oryctocephalus is a small to medium size trilobite with prominent eye ridges and four furrows connecting pairs of pits on its glabella. It has pleural spines, long genal spines and even spines on the pygidium. [x5]

Bathyuriscus is distinguished by its large forward-reaching glabella, pointed pleurae or pleurae with very short spines, and moderate pygidium with well-impressed furrows. Scaling up the largest pygidium found shows that a full grown individual may have reached seven centimetres in length, although complete specimens are smaller. [x1.5]

A moulted exoskeleton of *Bathyuriscus* (compare with image above and description on page 51). The free cheeks have been shed — the left one is rotated away, while the right one is inverted. This allowed the growing trilobite to escape its too-small exoskeleton and secrete a larger one. [x1.5]

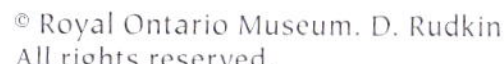

Kootenia is a medium-size trilobite with a large glabella and medium pygidium. It was originally distinguished from *Olenoides* by its lack of strong interpleural furrows on the pygidium. However, the expression of the furrows is variable, and *Olenoides* and *Kootenia* have recently been synonomized under *Olenoides*. [x1.5]

Trilobite Injuries

Trilobites were prey for other animals. Small fragments of the trilobite *Pagetia* occur in the digestive tract of an unnamed arthropod from the Middle Cambrian of China, and injured trilobites have been found in several Cambrian deposits. The *Olenoides* specimen from the Walcott Quarry (left) shows a well-defined injury on its left side. Its conspicuous W-shape suggests the wound may have been inflicted by *Anomalocaris* (compare with the *Anomalocaris* mouth on page 44). However no direct evidence, such as ingested trilobite fragments in *Anomalocaris* fossils, has been found to support this. The *Ogygopsis* fossil from the Mt. Stephen trilobite beds (right) shows a triangular injury to the right side of the animal. Healing, or scar tissue, at the bite margins indicates the animals were living at the time of injury, and managed to escape their attackers. They may have been in the vulnerable non-mineralized inter-moult stage when attacked.

Baby Trilobites

Juvenile trilobites are very small and are easy to overlook. The young *Oryctocephalus* from Mt. Stephen (right, with needle for scale) is one-eighth of adult size and has only five or six thoracic segments developed. Compare with the adult specimen on page 58.

Climate Change

*"Nested within a glacial age,
in turn within a greenhouse state,
the present interglacial stage is an
extremely unrepresentative sample
of past long-term climatic states."*

J.J. Veevers, 1990

Ice Ages are recurring events in Earth history. Rocks in Ontario and Quebec record the passing of great ice sheets over two billion years ago, and ancient sedimentary rocks of British Columbia's Windermere Supergroup tell of another ice age approximately 600 million years ago. Other long glaciations occurred in the Ordovician and Permian.

These glacial periods might be linked to plate tectonics. At all these times, the continents were massed together on the Earth's surface, producing a supercontinent. Supercontinents may set off ice ages by modifying weather conditions. On a large continent without access to intervening seas, little rain would fall inland. Large mid-latitude deserts, with their high reflectivity and heat loss, may have allowed winter snows to accumulate and ice caps to form. But the ice age that we know most about, the Pleistocene Epoch, began about 2.6 million years ago when the world map looked much as it does today. Continental massing cannot account for this glaciation.

In the 1920s Milutin Milankovitch revived James Croll's 1867 idea that climate is affected by changes in the amount of solar radiation reaching Earth's higher latitudes. This is controlled by three variations in Earth's orbit around the sun: 1) obliquity, the tilt of Earth's axis relative to its orbital plane, which varies on a 41,000 year cycle; 2) precession, the 22,000 year wobble in the axis orientation; and 3) eccentricity, the degree to which Earth's orbit strays from a circle, which varies over 100,000 years. Milankovitch predicted that glaciers should advance and retreat based on high latitude climate changes resulting from these cycles. Although the orbital variations are small, they can change the amount of solar radiation received at high latitudes by 15 per cent. Their effect on heat distribution over the globe may promote other changes in ocean and atmosphere circulation that bring on glacial epochs.

Measurement of ancient temperatures from oxygen isotope trapped in ice and sediment cores has refined our understanding of climate over the last glacial epoch.

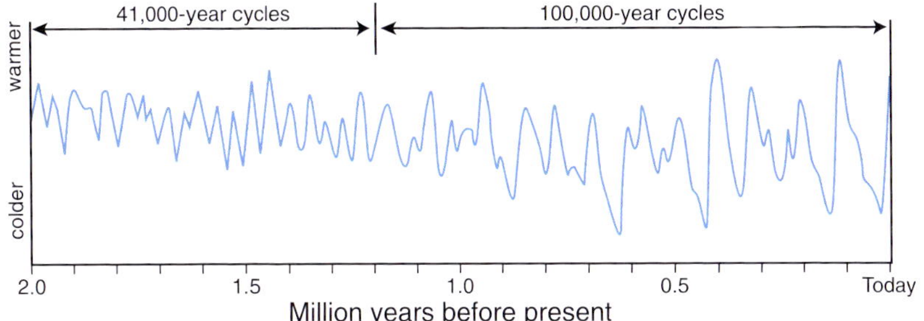

Temperature record from an equatorial Pacific deep sea core.

The Last 2,000,000 Years. There have been over sixty temperature cycles in the last two million years. (Not all of them are shown on the simplified graph.) Glacial periods were of longer duration than interglacials. In the older part of the record, glacial-interglacial pairs lasted about 50,000 years. About 1.2 million years ago the couplets began to last significantly longer—100,000 years or so—with glacial phases dominating much shorter interglacials. This is interpreted as the eccentricity cycle becoming more influential than the shorter period cycles.

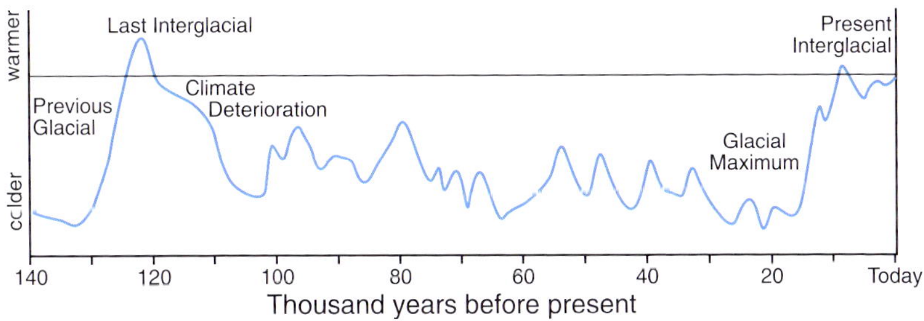

Temperature record from an Antarctic ice core, relative to 1950 (black line).

The Last 140,000 Years. A characteristic of interglacial periods is their rapid onset, followed by more gradual termination and descent into the next glacial period. The 140,000-year graph shows the abrupt end to the previous interglacial about 130,000 years ago and a brief respite of interglacial warmth before the climate began to deteriorate again. During most of an ice age conditions are less severe than at the glacial maximum. The most recent glacial period lasted about 100,000 years, with the coldest period occurring only 20,000 years ago. Within the overall glacial period there were a number of warm-cold oscillations.

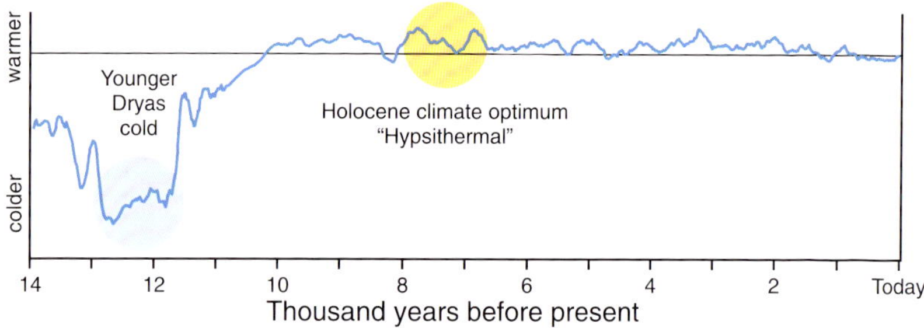

Temperature record from a Greenland ice core, relative to the 20th Century (black line).

The Last 14,000 Years. Just as the last glacial was ending, the Younger Dryas cold spell of the present interglacial period brought tundra conditions back to northern Europe from about 13,000 to 11,600 years ago. The Younger Dryas may have been caused by a disruption in the North Atlantic Ocean heat flow due to the release of massive amounts of cold fresh water into the ocean from the melting North American ice sheet.

Termination of the Younger Dryas was abrupt, with Greenland temperatures rising 10° C in a decade. Following recovery, temperatures over the last 10,000 years moved within narrower bounds. The temperature records available for this period indicate widespread regional climate variations not captured in the generalized graph.

A period of warm summers similar to today, called the climatic optimum or hypsithermal, occurred about 6,500 to 8,000 years ago in high latitudes of the Northern Hemisphere, but this was not a global or hemispheric phenomenon. Since then, temperatures up to the twentieth century were somewhat cooler.

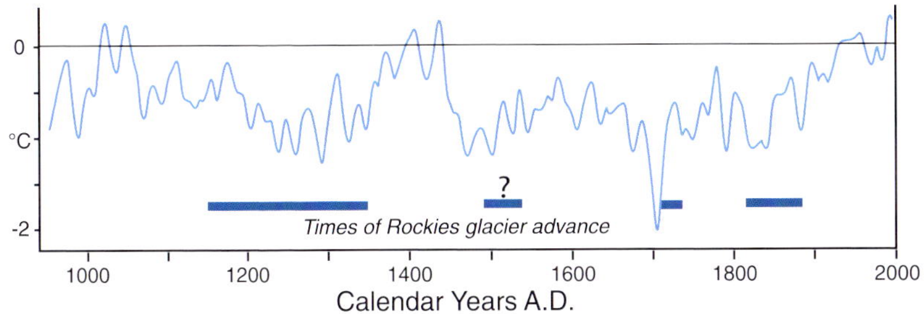

Summer temperatures in the Canadian Rockies relative to the period 1901–1980 (black line). (Luckman & Wilson, 2005).

The Last 1,000 Years. More recent periods have a greater number of sites and more diverse paleotemperature proxy records available, including ice cores, pollen, fossil shells, tree rings and cave deposits. The Northern Hemisphere records show regional temperature variations over the last millennium, making it hard to generalize paleotemperatures with a single graph. Past concepts of a global-scale Medieval Warm Period and a Little Ice Age have been abandoned in favour of a focus on regional changes.

Over the last two centuries explorers and photographers have documented the retreat of glaciers in the Rockies. The fronts of glaciers are still close to their terminal moraines (the gravel mounds that mark the limit of a glacier's advance). Geologists can determine the age of these moraines, and therefore, the year when the glacier was largest.

Brian Luckman at the University of Western Ontario has reconstructed the summer temperature curve for the Canadian Rockies. Glacier advances in the periods 1200 A.D. to 1370 A.D., the early eighteenth and the mid- to late-nineteenth centuries indicate the prevalence of cool summers. However, periods of warmth similar to modern times occurred in the eleventh and fifteenth centuries. The Yoho and Athabasca glaciers reached their greatest extents around 1844; the Columbia glacier in 1724.

Climate change has influenced the rise and fall of civilizations. Its effect on the hydrologic cycle, causing alternating periods of severe flood and drought, is thought to have contributed to the demise of the Mayan civilization in Mexico about 900 A.D. and the Hohokam culture in the southwestern United States after 1350 A.D.

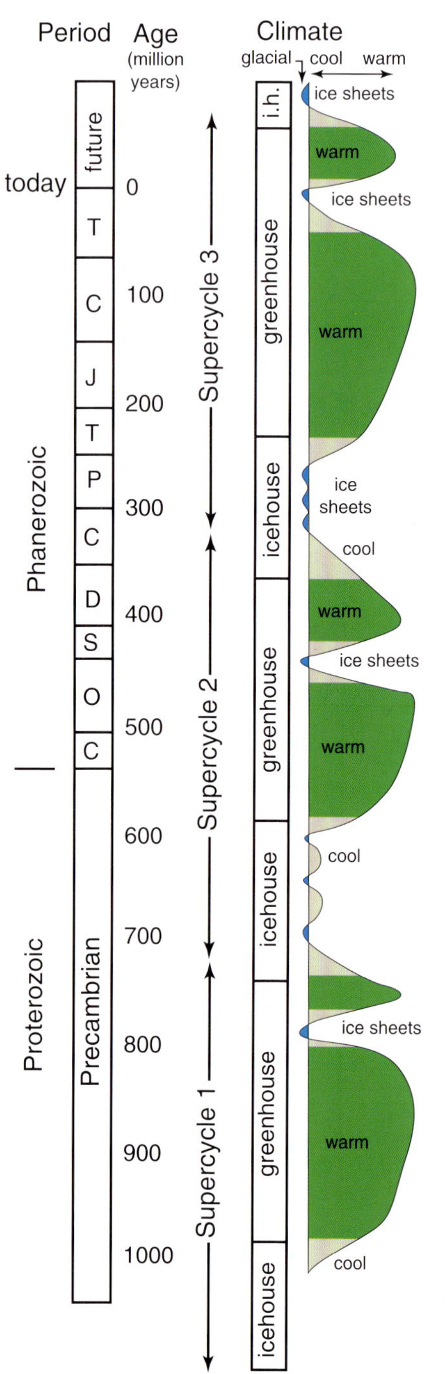

The Supercycle Concept

For most of the last billion years, Earth has been warmer than today. In 1990 J.J. Veevers, building on ideas of Alfred Fischer, compiled the climate curve at left. He recognized three *supercycles*, each lasting 400 million years and consisting of an icehouse-greenhouse pair.

Icehouse conditions are favoured when the continents are massed together, as in the late Precambrian Pannotia and Permian Pangea supercontinents. With the total length of mid-ocean ridges at a minimum, venting of CO_2 is low, the greenhouse effect is reduced, and continental weather patterns favour ice sheet formation.

Greenhouse climate prevails when supercontinents rift. The resulting long mid-ocean ridges vent much CO_2, and ocean waters warm, distributing heat to high latitudes.

Within each supercycle is an intra-greenhouse glacial. The cause for these interruptions is obscure. It may be related to a natural compensating mechanism such as increased cloud cover reflecting the sun's energy or a sink for CO_2.

Alpine Climate Change

Lake O'Hara and Opabin Lake are located sixteen kilometres southeast of the Burgess Shale, The varying type and quantity of pollen and conifer needles entombed in the lake-bottom muds provide a detailed record of alpine climate change. Cores of these muds make it possible to reconstruct the climate history of the past 10,000 years.

Following the end of the last major glacial advance 11,000 years ago, the climate in the Rockies began to warm. Ice left Lake O'Hara about 10,000 years ago. Shrubs and alpine tundra plants lined its shores but there were no trees. By 8,500 years ago, the area looked similar to today: the Opabin Glacier had retreated to the high slopes, shrubs and grasses surrounded Opabin Lake (265 metres above Lake O'Hara), and coniferous forest was well established around Lake O'Hara.

For 5,500 years the climate in the Lake O'Hara region was warmer than it is today. Between 8,500 and 3,000 years ago, timberline lay well above Opabin Lake and the Opabin Glacier may have disappeared altogether. The climate cooled again 3,000 years ago. The Opabin Glacier re-established itself and re-advanced to the shore of Opabin Lake while the timberline retreated to a region between the two lakes.

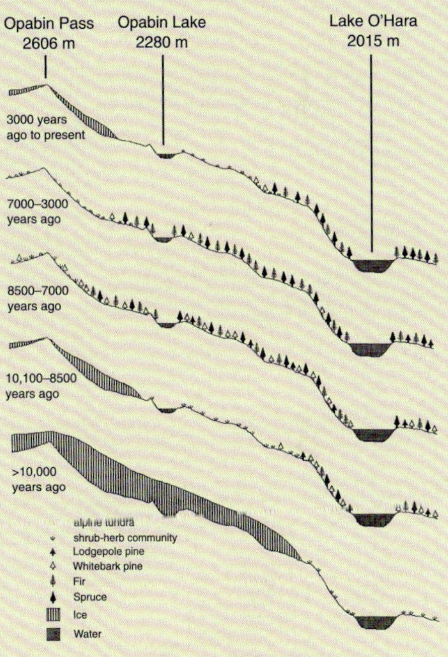

Temperatures began to warm again approximately 150 years ago. If present climate trends continue it is possible that the Opabin Glacier will disappear once more.

Weathering the Mountains

The glacial features of the President Range, seen across the Emerald Valley on the way to the Burgess Shale, show us how nature is wearing down the Rockies. Michael Peak is capped by the remnant of the glacier that carved out a bowl-shaped amphitheatre on its peak. This mountain feature is called a cirque. In front of the glacier are two arcuate gravel ridges. These are pieces of the glacier's terminal moraine, the rock debris pushed in front of the glacier as it advanced. A terminal moraine marks the maximum advance of a glacier.

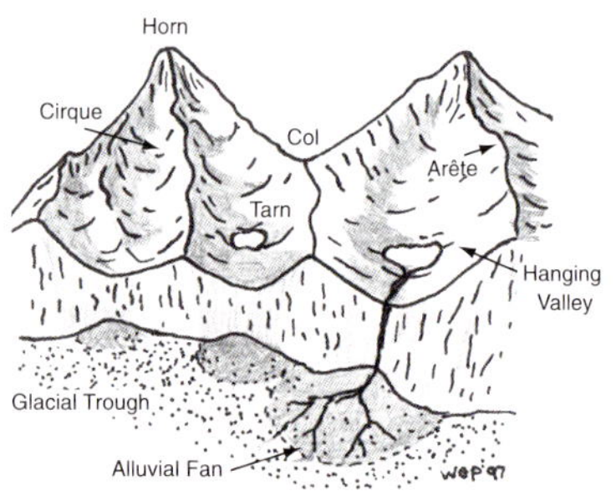

The dark patch on the east slope of the eastern moraine is rock slide debris from the arête of dark Eldon Formation limestone above it. The slide's debris ends abruptly at the cusp of the moraine, indicating that the glacier must have been present to block the falling rock. The rubble scar is far too sharply defined to have sat exposed to wind and water since the last Pleistocene advance 11,000 years ago. This moraine is probably less than 200 years old.

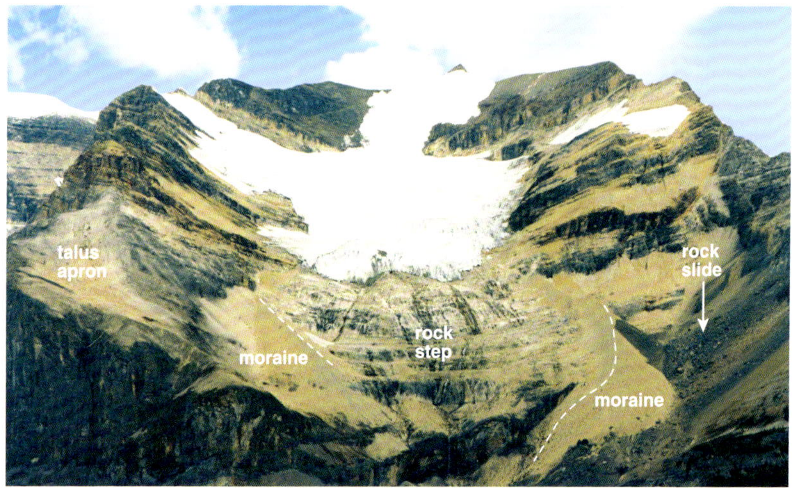

Glacial features on Michael Peak. The glacier has retreated into its cirque, revealing a rock step below, flanked by sharp-crested moraines marking its former extent.

Erosion is rapid in the Rockies today. Aprons of talus flank the cliffs below Michael Peak, evidence of repeated slides down the steep mountain side. The valley between Michael Peak and Mt. Carnarvon opens out into the Emerald Basin. At the junction, a large conical pile of sand and gravel called an alluvial fan. During spring runoff, torrents of water flow down the valley carrying vast amounts of frost-shattered talus and finer material. At the bottom of the basin, the slope of the stream flattens out and the water slows considerably. It can no longer carry its coarser sediments. Sands and gravels drop out, creating an alluvial fan, while finer silts and muds continue to flow toward Emerald Lake.

The striking blue-green colour of Emerald Lake is the result of sunlight reflecting off the suspended clay particles carried from the mountain tops. Deep water absorbs most light. If there were no sediment suspended in the water, the lake would appear dark blue to black. However, the suspended clays reflect light, but only the wavelength of light corresponding to the size of the particle. The tiny clay crystals reflect blue and green light.

Glacial outwash from the President Range is slowly building an alluvial fan in the Emerald Basin (lower centre). Ultimately Emerald Lake, glimpsed at lower left, will disappear as the valley is filled with debris.

Technical Literature

Geology

Aitken, J.D., 1997. Stratigraphy of the Middle Cambrian platformal succession, southern Rocky Mountains. Geol. Surv. Canada Bull. 398, pp.322.

Cook, D.G. 1975. Structure style influenced by lithofacies, Rocky Mountain Main Ranges, Alta.–B.C., Geol. Surv. Canada Bull 233.

Fletcher, T.P., and D. Collins, 1998. The Middle Cambrian Burgess Shale and its relationship to the Stephen Formation in the southern Canadian Rocky Mountains, Canadian Journal of Earth Sciences, v. 35, p. 413-436.

McIlreath, I.A., 1977. Accumulation of a Middle Cambrian deep-water limestone debris apron adjacent to a vertical, submarine carbonate escarpment, southern Rocky Mountains, Canada. In: Deep Water Carbonate Environments, H.E. Cook and P. Enos (eds.), Soc. Econ. Paleontol. and Mineral. Spec. Pub. 25, p. 113–124

Nesbitt, B.E. and K. Muehlenbachs, 1994. Paleohydrology of the Canadian Rockies and origins of brines, Pb-Zn deposits and dolomitization in the Western Canada Sedimentary Basin. Geology v. 22, p. 243–246.

Ney, C.S., 1954. Monarch and Kicking Horse Mines, Field, British Columbia. In: Fourth Ann. Field Conf. Guidebook, Alberta Soc. Petrol. Geol., Calgary, p. 119–136.

Stewart, W.D., 1991. Stratigraphy and sedimentology of the lower and middle Chancellor "Group," Rocky Mountain Ranges, southeastern British Columbia. Unpub. Ph.D. thesis, University of Ottawa, Ottawa Ontario.

Rasetti, F., 1951. Middle Cambrian stratigraphy and faunas of the Canadian Rocky Mountains. Smithsonian Miscellaneous Collections v. 116, no. 5. pp. 270.

Stewart, W.D., O.A. Dixon and B.R. Rust, 1993. Middle Cambrian carbonate platform collapse, southeastern Rocky Mountains. Geology v. 21 p. 687–690

Veevers, J.J., 1990. Tectonic-climatic supercycle in the billion-year plate-tectonic eon: Permian Pangean icehouse alternates with Cretaceous dispersed-continents greenhouse. Sedimentary Geology v. 68, p. 1-16.

Paleontology

Butterfield, N.J., 1990. A reassessment of the enigmatic Burgess Shale fossil *Wiwaxia corrugata* (Matthew) and its relationship to the polychaete *Canadia spinosa* Walcott, Paleobiology, v. 16, p. 287-303.

Caron, J.B., 2005. Taphonomy and community analysis of the Middle Cambrian Greater Phyllopod Bed, Burgess Shale. PhD thesis, University of Toronto, Toronto, 316 p.

Collins, Desmond, 1996. The "evolution" of *Anomalocaris* and its classification in the arthropod Class Dinocarida (nov.) and Order Radiodonta (nov.). J. Paleontology, v. 70, no. 2, p. 280–293.

Collins, D.H., D.E.G. Briggs and S. Conway Morris, 1983. New Burgess Shale fossil sites reveal Middle Cambrian faunal complex. Science v. 222, p. 163–167.

Eibye-Jacobsen, D., 2004. A reevaluation of *Wiwaxia* and the polychaetes of the Burgess Shale, Lethaia, v. 37, p. 317-335.

Fortey, R.A. and R.M. Owens, 1999. Feeding habits in trilobites. Palaeontology v. 42, part 3, p 429-465.

Ramsköld, L., 1992. The second leg row of *Hallucigenia* discovered. Lethaia v. 25, p. 221–224.

Walcott, C.D. (see separate list)

Whiteaves, J.F., 1892. Description of a new genus and species of phyllocarid crustacea from the Middle Cambrian of Mount Stephen, B.C. The Canadian Record of Science, v. V, no. 4, p. 205–208.

Whittington, H.B., 1971. Redescription of *Marrella splendens* (Trilobitoidea) from the Burgess Shale, Middle Cambrian, B.C. Geol. Surv. Canada, Bull. 209

Climate

Alley, R.B. 2000. The Younger Dryas cold interval as viewed from central Greenland. Quaternary Science Reviews v. 19, p. 213-226.

Jones, P.D. and M.E. Mann, 2004. Climate over past millennia. Rev. Geophys., v. 42, RG2002, doi:10.1029/2003RG000143

Luckman, B.H. and R.J.S. Wilson, 2005. Summer temperatures in the Canadian Rockies during the last millennium: a revised record. Climate Dynamics, v. 24, p. 131–144.

Petit, J.R. and others, 1999. Climate and atmospheric history of the past 420,000 years from the Vostok ice core in Antarctica. Nature, v. 399, p. 429–436.

Walcott's Publications

Charles Walcott filled five volumes of the Smithsonian Miscellaneous Collections with his papers on Cambrian Geology and Paleontology. Publications relevant to this guidebook are listed below.

Cambrian Brachiopoda: Descriptions of new genera and species. (1908) v. 53, no. 3.

Middle Cambrian Merostomata. (1911a) v. 57, no. 2.

Middle Cambrian Holothurians and Medusae. (1911b) v. 57, no. 3.

Middle Cambrian Annelids. (1911c) v. 57, no. 5.

Middle Cambrian Branchiopoda, Malacostraca, Trilobita and Merostomata. (1912) v. 57, no. 6.

Cambrian Trilobites. (1916) v. 64, no. 5.

Fauna of the Mount Whyte Formation. (1917) v. 67, no. 3.

Appendages of trilobites. (1918) v. 67, no. 4.

Middle Cambrian Algae. (1919) v. 67, no. 5.

Middle Cambrian Spongiae. (1920) v. 67, no. 6.

Notes on the structure of *Neolenus*. (1921) v. 67, no. 7.

Cambrian and Ozarkian Brachiopoda. (1924) v. 67, no. 9.

Pre-Devonian Paleozoic formations of the Cordilleran provinces of Canada. (1928) v. 75, no. 5.

Walcott also wrote a more popular account of his early work in the Rockies, titled *A Geologist's Paradise*, which appeared in the National Geographic Magazine (1911) v. 22, p. 509–521.

Further Reading

Beers, D., 1989. The Wonder of Yoho. Calgary: Rocky Mountain Books. *(Illustrated hiking guide to Yoho National Park, with historical anecdotes.)*

Briggs, D.E.G., D.H. Erwin, and F.J. Collier, 1994. The Fossils of the Burgess Shale. Washington, D.C.: Smithsonian Institution Press, 238 p. *(Likely the most accessible, authoritative and richly illustrated book on the Burgess Shale.)*

Conway Morris, S., 1998. Crucible of Creation. The Burgess Shale and the rise of animals. Oxford, Oxford University Press, 242 p.

Conway Morris, S. and H.B. Whittington, 1979. The animals of the Burgess Shale. Scientific American v. 241, p. 122–133.

________, 1985. Fossils of the Burgess Shale. A national treasure in Yoho National Park, British Columbia. Geological Survey of Canada, Misc. Report 43, 31 p. *(Contains many fossil illustrations. The original is out of print, but a scanned re-issue was released in 2003. Order at http://gsc.nrcan.gc.ca/bookstore/order_e.php)*

Gadd, Ben, 1986. Handbook of the Canadian Rockies. Jasper, Alberta; Corax Press. 876 p. *(Comprehensive guide to the geology and natural history of the Rockies.)*

Gould, S.J., 1989. Wonderful Life: The Burgess Shale and the Nature of History. New York: W.W. Norton & Co., 347 p.

________, 1994. The evolution of life on the Earth. Scientific American v. 271, no. 4, p. 84–91. *(Available on the internet at http://brembs.net/gould.html)*

Grove, J.M., 1988. The Little Ice Age. London: Methuen.

Nash, J.M., 1995. When life exploded. Time Magazine, v. 146, no. 23 (December 4, 1995), p. 38–46. *(Available on the internet at http://www.mc.maricopa.edu/dept/d10/asb/anthro2003/origins/life_explosion.html)*

Rutter, Nat, Murray Coppold and Dean Rokosh, 2006. Climate Change and Landscape in the Canadian Rockies. The Burgess Shale Geoscience Foundation, 130 p. *(An illustrated guide to the glacial landscapes of the Canadian Rockies. Includes a road log and site descriptions.)*

Yochelson, E.L. 1996. From Farmer-Laborer to Famous Leader: Charles D. Walcott (1850–1927). GSA Today, Jan. 1996, p. 8–9.

________, 1998. Charles Doolittle Walcott, Paleontologist. Ohio, Kent State University Press, 510 p. *(Covers Walcott's life up to 1907, before the Burgess Shale discovery.)*

________, 2001. Smithsonian Institution Secretary, Charles Doolittle Walcott. Ohio, Kent State University Press, 832 p. *(Covers the Burgess Shale years.)*

Geological Maps

Balkwill, H.R., R.A. Price, D.G. Cook and E.W. Mountjoy, 1980. Geology: Golden (East Half), British Columbia. Geological Survey of Canada Map 1496a, 1:50,000 scale.

Price, R.A., D.G. Cook, J.D. Aitken and E.W. Mountjoy, 1980. Geology: Lake Louise (West Half), British Columbia – Alberta. Geological Survey of Canada Map 1483a, 1:50,000 scale.

Price, R.A., D.G. Cook, J.D. Aitken and E.W. Mountjoy, 1980. Geology: Lake Louise (East Half), British Columbia – Alberta. Geological Survey of Canada Map 1482a, 1:50,000 scale.

Note. Maps 1483a and 1496a are essential for the Burgess Shale – Mt. Stephen area. Map 1482a, which covers Lake Louise, is useful for completeness.

Internet Resources

There are several sites dealing with the Burgess Shale available on the internet in 2006. Your first stop should be at The Burgess Shale Geoscience Foundation site (www.burgess-shale.bc.ca) which contains news, photographs and links to several other sites.

See also:

Yoho National Park:
www.pc.gc.ca/pn-np/bc/yoho/index_E.asp

The Burgess Shale: A Hidden Treasure In The Canadian Rockies:
park.org/Canada/Museum/burgessshale/titlen.html

Royal Ontario Museum
http://www.burgess-shale.rom.on.ca/en/index.php

Smithsonian Institution:
http://paleobiology.si.edu/geotime/main/index.html

Peabody Museum of Natural History, Yale University:
www.yale.edu/ypmip/locations/burgess/

University of California Museum of Paleontology:
www.ucmp.berkeley.edu/cambrian/burgess.html

Animated fossil reconstructions:
homepage1.nifty.com/burgess/aa.html

Sources

Plate tectonics diagram (p. 6): Wayne Powell. Cambrian World map (p. 12): PALEOMAP Project, C.R. Scotese, used with permission. Stratigraphic column (p. 15) after W.D. Stewart (1991). Charles Walcott portrait (p. 23): Smithsonian Institution Archives photo 84-16281. Helena Walcott (p. 24): Smithsonian Institution Archives photo 85-11428. Arthur Brown (p. 24): Charles Walcott, Smithsonian Institution Archives photo 31656. All Smithsonian photos from Record Unit 95, Photograph Collection 1850s-, and used with permission. Burgess Shale Quarries (p. 25): Wally Randall. Cathedral Escarpment reconstruction (p. 27): Geological Survey of Canada (GSC) (adapted from Conway Morris and Whittington, 1985), used with permission. Artist's reconstruction of the Burgess Shale mudslide (p. 27): copyright Smithsonian Institution, Mary Parrish, all rights reserved. Walcott Quarry (p. 32): Smithsonian Institution Archives photo 44682. Walcott Quarry panorama (p. 33): Cherry Burton. Artist's conceptions of soft-bodied fossils (except where noted): Thomas Saunders. Paintings of *Hallucigenia* (p. 36), *Opabinia* (P. 43), *Anomalocaris* (p. 44) and *Pikaia* (p. 47): copyright Smithsonian Institution, Mary Parrish, all rights reserved. Portrait of Franco Rasetti (p. 55): The Ferdinand Hamburger Archives of The Johns Hopkins University, used with permission. Paleoclimate graphs : 2 million years (p. 63): oxygen isotope data from various deep sea cores from the Ocean Drilling Project, Texas A&M University. 140,000 years (p. 63): after J.R. Petit and others, 1999. 14,000 years (p. 64): after Alley, R.B. 2000. 1,000 years Canadian Rockies (p. 65): after Luckman and Wilson, 2005. Alpine Climate Change (p. 67): Mel Reasoner. Glacial landscapes (p. 68): Wayne Powell.

Photographs of fossils:

Aysheaia (p. 35), *Opabinia* (p. 43) and *Pikaia* (p. 47, upper): Courtesy of the GSC and with permission of the Minister of Public Works and Government Services.

Hallucigenia (p. 36), *Marrella* (p. 37), *Burgessochaeta* (p. 37), *Waptia* (p. 38, left), *Burgessia* (p. 39), *Sidneyia* (p. 43, lower), *Leanchoilia* (p. 40), *Yohoia* (p. 41), *Wiwaxia* (p. 42), *Anomalocaris* claws and mouth (p. 44), *Anomalocaris* (p. 45), *Pikaia* (p. 47, middle) and *Olenoides* (p. 52): copyright Royal Ontario Museum, J-B. Caron, all rights reserved. *Naraoia* (p. 46): copyright Royal Ontario Museum, D. Rudkin, all rights reserved.

Waptia (p. 34 and 38, right), *Burgessochaeta* and *Burgessia* (p. 34), *Canadaspis* (p. 38), *Sidneyia* (p. 39 middle) and *Pikaia* (p. 47, upper): Charles Walcott.

All trilobite photographs (except as noted): copyright Royal Ontario Museum, D. Rudkin, all rights reserved. Trilobite slab (p. 50): Lisa Holmstrom.

All other photographs and diagrams by Murray Coppold.

Acknowledgements

The authors are grateful to several individuals for their help in producing this book. Jean-Bernard Caron and David Rudkin, Royal Ontario Museum, provided fossil images of the Burgess fauna and trilobites. They gave valuable editorial assistance to the second edition, sharing the results of their research that led to the updating and revision of several sections. Responsibility for errors remains with the authors.

Randle Robertson, Executive Director of The Burgess Shale Geoscience Foundation, has tirelessly promoted education about the Burgess Shale and its message for over a decade. Desmond Collins, Royal Ontario Museum, gave consistent support to this project through the years. Mary Parrish, Smithsonian Institution, graciously contributed her paintings of the Burgess fauna. Ellis Yochelson, Smithsonian Institution, provided insights into the life of Charles Walcott and the Walcott family. David Sargent, Geological Survey of Canada, supplied the GSC image of *Aysheaia*.